Journey Back To Eternity

THE STORY OF CREATION

by Jack Cobleigh

ISBN: 0-7596-7685-2

This book is printed on acid free paper.

1stBooks - rev. 04/29/02

The Table Of Contents

Acknowledgments

I would be remiss and ungracious if I did not acknowledge the many people who assisted in the completion of this book. Without their help, this project could never have begun, let alone completed. The continued encouragement of my friends and family bolstered my spirit to complete the work. To these many dear people, I would like to offer my thanks and heart felt appreciation for their assistance.

Since my experience in computers was only in Computer Aided Design (CAD), I had some interesting surprises awaiting me. I want to thank my stepson, Jim Dale, for making personal computers available to us, and for setting up all that complicated equipment. These computers were wonderful toys for our grandchildren. It was a little embarrassing, when the six-year olds asked me to turn one on, and the eight-year olds had to show me how.

I want to thank my brother and sister-in-law, Rev. Jerry and Kay Cobleigh, for suggesting that I write the book. For many years, I have collected many interesting facts about ancient, Middle East history, cultures and civilizations, archaeology, and Bible prophecy. I have filled four notebooks with such data. The surprising suggestion, of not just sharing this information with the classes that I teach but to put it in a book for many people to share in, was a great challenge for me.

Thanks to Jerry and Kay, the seed was planted, but what do I do now? My stepson and his wife, John and Jessica Dale, graciously and patiently, introduced me to the wonderful world of word processing. I charged ahead confidently, until I found myself in deep trouble.

I called upon dear friends of ours, Terry and Jean Babics, to help unsnarl the electronic entanglements that no one had ever heard of before. Terry and Jean have been very helpful in the installation of new equipment and adding new programs. I truly appreciate the inexhaustible supply of technical knowledge from our family and friends. Without the help of these dear people, this project could not have been completed.

A special thank you goes to Miriam Walton, who graciously consented to edit this book. Miriam and her husband, Ken, have been wonderful friends of ours for many years. I truly appreciate her wonderful work in catching all of the missing commas, correcting the strange sounding adverbs, and pasting together the split infinitives.

A special, warm, heart-felt appreciation goes to my lovely wife, Yvonne, who faithfully supported me throughout the project, and her quiet patience, sound advice, and wonderful insight. Yvonne, as always, has been a special blessing to me. In humble gratitude, I give thanks and appreciation to the LORD, for His continued blessings, of glorious inspiration and revelation.

May God's Richest Blessings be upon ALL of you

In Great Appreciation, Always

Jack Cobleigh

Introduction

Throughout the ages, man has sought to know the answers to basic questions of his own existence. How did the universe come into existence? How did all life begin? Why was I born, and why was I born at this time in history? Is there a reason for my life? Am I part of some great grand plan and purpose for life, or is this all there is? Throughout history, men, of philosophy, history, theology, and science have sought to resolve these questions. Even primitive cultures and civilizations have offered up traditions and legends of their origins. Perhaps, you have asked these very same questions, but have never found any answers that completely satisfy your quest for truth.

You are cordially invited to join me in the exploration of these serious questions. Together, we will explore the most prominent and trusted theories of today. The theories of *Evolution* and *Creation* will be thoroughly discussed, with their many variations. Perhaps your own theory will be among the topics reviewed. I challenge you to travel with me, to search out the truth about the theories of man, as well as the claims of the only one ever to give an eye witness account of the creation of all things: God. All Scriptural references in this book are from the Scofield Reference Bible, (KJV), the 1967 Edition.

Within the pages of the Bible, God relates to us not only how all things came into existence, but also who created them. We shall delve deeply into the most popular theories of men, both for *Evolution* and for *Creation*. We shall also explore what God has to say in His Word. In fact, we shall examine and follow every word of God's claims every step of the way. We shall travel within our imagination back through time, and beyond. We shall *"Journey Back to Eternity,"* to be in the presence of Almighty God, at a time when nothing else existed. Together, we shall experience the creation of God's Heaven in its entire splendor. The Throne of God and the Temple of God are created upon a crystal sea. The choruses of majestic angels resound through the corridors of Heaven, and we are there to rejoice with them.

We can witness the universe being created, in all of its dazzling brilliance. Astonished by the size of the stars, their blazing colors, and brightness, we will be awed by the beauty and vastness of the universe. As we journey together, we shall be present at the creation of planet Earth, and witness the six days of Creation, ending with the creation of Man. Let your imagination and senses behold the exotic Garden of Eden in its pristine holiness. You are there when man falls, and is cast out of the Garden. We experience every detail as the Creator God's glorious plan of *Creation and Redemption* unfolds before our eyes. The quest for resolution to the hard questions of existence shall be rewarded, as we *"Journey Back to Eternity."*

Chapter 1

In the Beginning…

"In the beginning" is a term that has brought wonder and awe to the hearts and minds of mankind ever since he walked this earth. Man has searched the heavens above and deep within his own soul to find answers about his very existence. Why am I here? Why was I born at this time of man's history? Is there a purpose or plan for my life? Did I evolve from some ancient primordial slime, as some scientists say, or was I created by an all knowing, all powerful, all loving God? These are just a few of the many questions I have asked myself as I am sure you have asked yourself, as we have pondered these things. Let us explore some of these awesome questions and as we pursue and examine them together, perhaps we will find illumination.

There are many theories that have been put forth about how the universe was created and how man came into being. Many of these theories are very ancient but some are of recent vintage. Many are quite humorous such as the ancient eastern idea that earth travels through the heavens on the back of a turtle. Let us pursue some of the more modern concepts and not get entangled with the ancient or we just may fall off the edge of the "flat earth." What prevails today is the theory of Creation and the theory of Evolution. These two concepts are at opposite poles from one another and are bitterly contested by their proponents. I shall not presume to prove or disprove any theories by way of scientific means but I would like to examine and illuminate various ideas and perhaps come to some conclusions.

The theory of Evolution had its beginnings in the 19th century when Charles Darwin wrote his book *"The Origin of Species.."* He did not attempt to publish his book until 1859, very late in his life, because he felt his findings were too controversial and his conclusions were unprovable. When it was finally published it was met with great resistance, but it did attract a number of ardent supporters. The concepts that Darwin put forth were soon expanded to include new theories on how the universe itself began. It is commonly known as the *"Big Bang Theory"* and this theory has grown to become accepted as truth as has the theory of Evolution.

Liberal colleges and universities have been teaching this theory as well as the theory of Evolution as though it was absolute truth and they have dropped all notions that it is only a theory. Generations of students have been exposed to these teachings; and now a Creator God has been replaced by science, chance, and nature. Let's take a little closer look at these theories, not from the standpoint of the scientist, but just from the viewpoint of the average lay person using only logic and common sense. We will draw our own conclusions as we review some very interesting information that is rarely ever addressed.

I THE BIG BANG THEORY

I would like to start with the theory of the "Big Bang" the so-called beginning of the universe. Scientists today have supposedly calculated to the exact nano-second what took place at the formation of the universe. All of the "cosmic dust" in space began to draw closer and closer together by way of electro-magnetic forces. As all of this space material was gathered together, the magnetic pull upon each electron became stronger and stronger and the colliding electrons created heat. The swirling mass of material grew more and more compact and dense and was rapidly shrinking into a massive fireball with an astronomical amount of heat and pressure. These tremendous forces were increased at such phenomenal rates that this huge galactic fireball suddenly exploded with an incalculable great force and scattered this fiery matter throughout space which ultimately became the solar systems of suns, moons, planets, and asteroids. This highly charged fiery material changed into all of the elements that compose gasses, minerals, and metals that are known today. Scientists also tell us that 99% of all the material within our own solar system is in the sun. Most scientists believe everything that is in the heavens and in the earth were "created" by the Big Bang. The biggest problem with this particular theory is that no scientist or theoretician has ever attempted to explain where the enormous amount of material came from to create the billions of galaxies that fill the universe. Perhaps all the data on electro-magnetic forces, on the heat and gravity, etc. is accurate and all of the computer models are correct, but the nagging question still remains. Where did all of the matter come from? The common sense question is always there. If you claim, there is no Creator God, then you must be prepared to explain how and why everything was created and for what purpose does life exist?

Since eternity only exists in the spiritual realm, then all other matter must, by its very nature, exist within the space/time continuum. All matter must be subject to time and space and to all of the physical laws that govern them. If all matter must be subject to these laws, then it is impossible for matter to independently create itself.

Today, science, the news media, the entertainment media, the educational system, and the liberal religions of our great country seem to be willing accomplices in promoting these Godless, humanistic theories of evolution without accepting responsibility for their decisions. Generations of children have been taught that about 4.5 billion years ago our universe exploded into existence and that life began out of the slime and ooze of this earth. Creation by mindless matter is quite a tale to sell to the public and to the scientific community, but it has been done very successfully. When I was a boy, the time of "universal creation" was about 1.5 billion years ago. As more and more theories of creation were added to the list, the only proof that was required was to add 100 million or a billion more years to the time frame for it to all happen. It didn't seem to matter how ridiculous or radical the proposition was, just add another billion years for it to occur, then everyone will be happy and accept the theory as truth. This is the level to which most scientists today have sunk, accepting it all in total, blind faith.

II THE THEORY OF EVOLUTION

The theory of Evolution is a very complicated theory covering many different disciplines of science. These disciplines include Biology, Genetics, Geology, Oceanography, Anthropology, Physics, and many other scientific fields. There is a vast difference between the theories that Darwin originally proposed and the science of today, because of the great strides that science has taken in the last one hundred years particularly in the field of Genetics. Many of the neo-Darwinists of today are totally convinced that the shaky, implausible theories of Darwin are absolute facts in spite of the enormous lack of scientific data that is required to support any theoretical proposition. The standard protocol of any establishment of scientific facts is to first build a foundation upon empirical data to prove the worthiness of any theory. When dealing with evolution, on the other hand, most scientists seem willing to ignore required scientific protocols and procedures to avoid facing the alternative model of the causal forces of creation; and that alternative is an Almighty Creator God. There just is no middle ground, either this universe created itself and the life in it, or it was all designed and planned by a Supreme Creator. The scientific world that adheres to the tenets of evolution are now faced with the fact that thousands of absolute formulae, equations, and established scientific procedures are all founded upon the result of random chance, chaos, and accident. It certainly requires a great deal of faith to believe something that is based on accident and chance. It seems logical to me that accepting anything in blind faith is, at best, an irrational choice and most certainly unscientific.

Let us take an abbreviated look at Darwin's theories to get a more accurate picture of his concepts. In his classic book *The Origin of Species*, Darwin argued three important related propositions. The first was that *"the species are not immutable."* By this he meant that new species have appeared during the course of earth's history by a natural process called *"descent with modification."* The second proposition was that this evolutionary process could be extended to account for all or nearly all the diversity of life, because *"all living things descended from a very small number of common ancestors, perhaps a single microscopic ancestor."* The third proposition, and the most distinctive to Darwinism, was that this vast process was guided by natural selection or *"survival of the fittest."* These three propositions are the very core of Darwin's evolution theories.

Let's examine Darwin's first statement *"the species are not immutable,"* in other words species will not always reproduce *"kind after it's own kind"* or *"seed after it's own seed."* He further proposes that somewhere in time a species will produce something other than it's own kind a *"variation"* as Darwin called it, and what scientists today call *"mutation."* Mutations as a source of "new species" seem to be at polar opposites of *"the survival of the fittest."* The mutant deformed offspring of any species is always the first to die by the parent allowing it to die and or by becoming a meal in the food chain. This is the natural way of things. The weak, the sickly, the mutants all perish and do not survive a single generation, let alone hundreds of generations of the "unfittest" to propagate a new species. Of course, if you extend the time frame to hundreds of millions of years, then anything is possible because there are no scientific models to prove such an outlandish proposition. Therefore, one must accept it totally on blind faith.

The second of the propositions is complimentary to the first *"all living things descend from a few or even one microscopic ancestor."* If one can believe that out of the enormous variety of species of plant and animal life came the mutations of only a few "parent stock" species, then it is not too hard to believe that all species (all life) came from a common Protozoan, one-celled microbe. Where did the single-celled microbe come from is the next logical question? If there was no other life on earth, then how did this single life form appear? Darwinists claim that within some ancient pond of primordial soup of minerals and gasses mixed together to form amino acids, the "building blocks of life," and randomly selected all of the correct RNA, DNA, and the Genetic coding for ALL life forms that will ever be. This is absolutely amazing! The arrogance of trained scientists postulating such irresponsible proposals boggles the mind and offends ones sense of logic and reason. Not at all, they might reply, because this didn't occur in tens of millions of years, but it took billions of years to happen. How can one dispute that sort of scientific proof?

Science today, with all of its' vast array of hi-tech equipment and knowledge even with super-computers cannot duplicate the conditions to create life. Not even one little single-celled protozoa has ever been created in the laboratory. Scientists have, for many years, attempted to create life, but without success. They cannot create in the lab with all of their combined knowledge the very thing that they are asking us to believe occurred on its own by random chance.

The third concept of Darwin's trilogy is *"the survival of the fittest"* or sometimes called *"natural selection."* This concept covers the entire vast range of life upon this planet. This theory emphasizes that each specie must struggle to survive or face extinction. Each individual animal or plant must be strong and healthy to survive and must pass on this strength to the next generation for its specie to survive. This does occur in nature, but it seems that the Darwinists want to add further stipulations to this precept. They claim that the "fittest" may have helpful genetic mutations that are passed along generation after generation to create many new and different species over vast periods of time. As we have seen earlier, mutations do not survive.

Experiments have been conducted with "hybrid" animals that have been turned loose in the wild to see if they could reproduce on their own or better yet produce a new species. To the utter disappointment of these hopeful scientists, not only did they not create any new species or even reproduce their own "hybrid" man-made state, but also they reverted back to their original basic stock. What a resounding defeat for the evolutionists this was. All of their hopes were tied up in something that could be proven via some kind of controlled experiment that they could point to that would support their claims. This, I suppose, would be a good place to add another fifty million years to the project to make it work. Evolution does not work either in the lab or in nature, but that does not seem to bother the staunch Darwin followers. The only option left to them is that this beautifully designed universe is maintained and regulated by a supernatural force, a Supreme Being, a Creator God.

THE LAWS OF THERMODYNAMICS

There is actually a considerable body of sound scientific evidence that contradicts the theory of evolution, some of which appears to be absolutely incompatible with the theory. There are laws of Physics, Chemistry, Biology and many other disciplines that are totally ignored or are severely twisted to accommodate their particular philosophy of the origin of the universe and the origin of the diversity of life.

The theory of evolution violates two of the most fundamental laws of nature: *the First and Second Laws of Thermodynamics*. The First Law states that "nothing now can be either created or destroyed." The Second Law states "every change which takes place naturally and spontaneously tends to go from a state of order to one of disorder." Darwin's theories are at complete opposite ends of the spectrum with these two laws. Evolutionists are forced to totally ignore certain laws that do not fit their philosophy or take a stand and declare that the fundamental laws of nature apply to everything else in the universe except for theories of evolution. Let's examine The First Law of Thermodynamics, it states that "nothing now can be created or destroyed." No matter what changes may occur in nature, the amount of matter and energy remains constant.

"Nothing now can be created or destroyed" means that everything that exists was at one time created, not evolved. It is impossible to create anything new or destroy anything out of existence. Many transformations may take place, but the total amount of matter and energy will always remain constant. The Second Law of Thermodynamics states "every change which takes place naturally and spontaneously tends to go from order to disorder." This means that it is impossible for any species to recreate itself to a higher order, such as an ameba into a reptile, or grass into a redwood tree, or an ape into a man. Darwinists cannot escape the laws that govern this universe and the life upon this planet.

In chemistry, amino acids do not spontaneously combine to form proteins, but proteins spontaneously break down to amino acids and the amino acids slowly break down into simpler chemical compounds. This is just the opposite of what must occur to create life, according to Darwin. Even the Alchemists that dream of turning lead into gold must bow to the "Laws of Entropy," as they are often called. Even with the space-age technology that is available today such as super-computers, cyclotron atom smashers, electron microscopes, etc., scientists still cannot perform a simple task of "evolving" the basic molecular structure of lead to that of gold. No matter how many times these Laws of Thermodynamics have been tested or by what discipline, mathematics, chemistry, biology, physics, etc., they have always proven to be valid. It must be extremely frustrating to be totally convinced about something so important to your very existence and the purpose for your life, and then discover the foundation of your beliefs is built on quicksand. Theories that proclaim you are evolved cannot be proven and whenever these theories are put to the test they fail miserably every time.

A great number of scientists have abandoned the Evolution model in favor of the Creation model. In spite of the tremendous losses of evolution devotees from within the scientific community, and in public support, the resolve of those who propagate these theories of evolution has not diminished, particularly in the fields of education and the entertainment media.

III CREATION vs. EVOLUTION

It was a very popular indoor sport in the 1980's for scientists from the Institute for Creation Research in San Diego, California or from the Creation Research Society to engage in debates on college campuses across the country with scientists who were experts within their fields of evolution. The campus auditoriums were always filled to capacity. The news media were often in attendance, because this was an extremely unusual event that anyone would dare to challenge the long established "facts of evolution." The students and the general public attending these debates had never before heard or read any information on creation, because of the total censorship of any data that supports creation in the textbooks or in the lecture halls. Evolutionists within the educational systems along with powerful liberal teachers' unions have established what will appear in every textbook from kindergarten to university level education.

Every nature show televised, such as "National Geographic," "Nova," or "Nature," and a host of others all emphasize the humanistic evolutionary philosophies. The audiences that witnessed these debates were astonished to hear how utterly shaky the foundation of evolution really is and how strong the position is for creation. Often using the very same data as the evolution model, the creation scientists would always present a more logical and solid presentation. The format of utilizing the same data and presenting a much stronger case was shocking to the students, because they are now asking "why wasn't this information made available to them so that they could make up their own minds as to which of the two models was correct?" The most condemning portions of the debates were when data and theories of the evolution scientists were challenged and proven wrong, and then provable working theories were introduced for the creation model.

These Creation vs. Evolution debates received big news coverage for awhile until the evolutionists realized that they were being badly mauled in these debates. The camp of the evolutionists was ravaged and as quickly as these debates started they seemed to disappear after a few years. At least the news coverage stopped abruptly. If the debates still continue, no one really knows because the general public has been kept in the dark. Nevertheless, a number of scientists who were once heavily committed evolutionists are now thoroughly convinced that the universe and its' infinite variety of life was created by an all-loving and all-powerful Creator God. The universe and all that is in it is a great grand scale design that by the simplest logic demands that there must be a Great Designer. Scientists from every discipline, every field of specialization, are now realizing that the foundation of evolution is like a giant house of cards. If its basic structure is deeply flawed, then everything that was built upon it must also be flawed. A belief system must be built upon truth; otherwise it is only wishful thinking. Yet the followers of evolution seem to be content that it is a viable belief system and pursue it with unquestioning enthusiasm. After all, they were taught the tenets of evolution in every phase of their education.

Many scientists would proudly refer to the fossil record as proof of their belief in evolution. Then let's examine the fossil record, again not from a scientific approach, but from the viewpoint of a logical, reasoning layman. There are many others who would gladly take up the challenge on the scientific level. These scientists, in the past, have presented their case with empirical proof and sound data, but it was like "a voice crying in the wilderness." Today is quite different; scientists are taking a brand new look at the evidence. So let's take a look for ourselves.

THE FOSSIL RECORD

Darwin's theory claims that *"species are not immutable"* or in more simple terms, one specie can produce a whole new specie. This idea is the very core of the theory of evolution. If given enough time, perhaps 10 million years or so, a specie could mutate into an entirely new specie. The evolutionists believe that by the process of mutation, the vast array of life appeared upon earth. This theory now presents a very large problem…the fossil record. The fossil record is a huge burden that evolutionists must carry, because without this empirical proof, they have no case.

What is this process of mutation and how does it affect the fossil record? To simplify things, let's reduce 10 million years of mutation into a short formula. Specie (A) will mutate into a new strain of itself, mutation (B), which will then mutate into a new specie (C), thus a brand new species is created.

Specie (A) to (AB) to mutation (B) to (BC) to new Specie (C)

Between specie (A) and new specie (C) there are 10 million years of mutations of (AB), (B), and (BC). Let's be more specific. Some scientists believe that birds evolved from reptiles, others believe that they evolved from dinosaurs. We will use the reptile to bird model for our example.

If reptile (A) was walking along one day and brushed up against a rock and knocked off a scale or a piece of hide and a feather began to grow in its' place, this would be the beginning of the mutation process. Thousands, or hundreds of thousands of generations later, more and more feathers appear and the body shape would begin to deform to accommodate the awkward feathers. A few hundred thousand more generations pass and now reptile (A) has evolved to the mutation (B) level, half reptile and half bird. Again, we travel through millennia after millennia and a few hundred thousand more generations of mutations and we see a brand new specie called a bird (C). Bird (C) now is covered with feathers, has a new body with new organs, and has the ability to fly. This is essentially the base core of evolution.

Therein lies the great problem for evolutionists. If Reptile (A) mutated over 10 million years to Bird (C), then there should be a vast fossil record of the 10 million years of mutations. Various stages of mutations should be easily found, including fossils of half reptile and half bird. In fact, the vast majority of ALL fossils found should be in some stage of the mutation process. No fossils have ever been found that reflects any stage of mutation, or half of one specie and half of another. This is the baggage that evolutionists must live with, a difficult premise to believe in to start with, but also there is absolutely no empirical proof to back up such claims for the origins of life. While we are discussing the many problems of evolutionary theory in the archaeological realm, let's expand our examination to the "geological column."

THE GEOLOGICAL COLUMN

The *"geological column"* is a term used by scientists to describe the various strata or layers of the earth's mantle. The earth's crust is layered in a variety of geological rock formations. According to the tenets of evolution, all changes affecting the earth and its' life forms took place very gradually over extremely long periods of time. Therefore, there could not be any instantaneous sovereign act of creation, or any "catastrophic event" such as a worldwide flood that would dramatically alter the earth and the life upon it.

The "geological column" theory states that the very oldest of the rock formations are at the very bottom layers and each successive age would leave its' own distinct fossil record layer after layer until at the top layer there would be the newest formations. After all, it takes hundreds of millions of years for these ages to pass and for a new evolutionary age to begin. This would appear to be a logical assumption if the earth is 4.5 billion years old and that the elements and life itself are in the process of creating themselves. For the evolutionists, there can be no catastrophic independent creative force. There can be no Creator God to speak the universe into existence. There can be no Supreme Being that could catastrophically intervene into the affairs of this world, such as a worldwide flood. The world must evolve by itself, slowly changing itself age by age, eon after eon, and depositing its' rock and fossil record all neatly layered one after the other. The fossil record in the geological column should, therefore, be neatly layered with the oldest record on the bottom and the youngest record at the top. This is the theory put forth by the followers of the evolution model, but again the facts do not support the theory. The truth is that the geological and archaeological excavations rarely if ever find the "geological column" in this prescribed order. What is the truth about the reality of excavations? Does this "geological column" actually exist in the manner in which we have been taught? Let's examine some of these finds.

DINOSAURS AND MAN

Everyone is fascinated with the discovery, in this century, of the now extinct dinosaurs. Dinosaurs have excited our imaginations, from the very old to the very young. According to the precepts of evolution, these large and unusual creatures became extinct about 65 million years ago, during what is called the Jurassic Age. This is a rather recent event in the history of 4.5 billion years of age for planet earth. The reality is that 65 million years is an extremely long time. If these animals lived 65 million years ago, as evolution scientists tell us, then the fossil record in the "geological column" should be buried very deep under the many following ages. The truth is, those dinosaur fossils are found very close to the surface of the digs. Most of the dinosaur bones that have been discovered in the United States have been found in the hills of Montana, Wyoming, and Colorado. Hippopotamus and giraffe bones have been found with those of ocean creatures in plains of Nebraska. These fossils have all been discovered very near ground level. This is consistent with the finds in Alberta, Canada; Europe; Mongolia; and in Russia as well. The fossil record does speak for itself and states emphatically that the interpretation of the fossil record and of the "geological column" has been falsified to support an invalid premise. What we are never told is that many of the "dinosaur bones" that are discovered are NOT fossilized which definitely proves that extinction in the so-called Jurassic Age occurred a very short time ago, perhaps during a global flood.

Another amazing problem for evolution scientists is in a little town in Texas, called Glen Rose. In the Dinosaur Valley State Park, in Glen Rose, Texas, are many tracks of dinosaurs along the Palaxy River and in the river as well. Dinosaur tracks at ground level and well-preserved are an exciting find to be sure, but they aren't the only tracks there. There are tracks of a man who was walking along, then turned left for a few strides, then he knelt down and placed a perfect handprint in the mud. Afterwards a dinosaur walked across his path and left three imprints on top of those of the human footprints.

At the Glen Rose site, there are 203 dinosaur tracks and 57 human footprints, and seven of their tracks overlap one another. A stone hammer with a wooden handle was found encased in stone, as well as a bone from a human finger. All of these finds were discovered in the same Cretaceous limestone layer. Similar finds of tracks with dinosaurs and man together have also been discovered in Russia and in Australia as well. Haven't we been told that this cannot happen? Here are absolute empirical proofs that man and dinosaurs lived at the same time. "Don't bother me with the facts, I have my own theory" is generally the response to such proof.

They do agree that the extinction of the dinosaurs was a sudden event. There have been many theories as to how they died off. Some believe it was a meteoric collision with the earth that caused a global dust cloud that blocked out the sun for years and created a suffocating unbreathable atmosphere. Others believe that it was global warming, or global cooling, or global disease, that caused the extinction of the dinosaurs. Any radical, implausible proposal would deserve serious consideration except the proposal of an Almighty God causing a global flood in righteous judgment upon a wicked and violent world.

Who can forget that great picture in the newspapers, in 1977, of the Japanese fishing trawler that caught an enormous, decaying, sea monster? It was believed to belong to the Plesiosaur group. This was a marine reptile from the age of the dinosaurs that had recently died and was in the process of decay. This stirs one's sense of reason. If the dinosaurs died out 65 million years ago, then where did this dinosaur come from, and are there any more like it in the seas?

Could such a global catastrophe have really happened and would the fossil record support such an event that has been proclaimed in the Holy Bible and by many scientists throughout the world? If such an event did occur, then the fossil record would abundantly reflect it. Are we being told the truth about the "geological column," or the fossil record, by the evolutionists?

THE CATASTROPHIC FLOOD

The scientific record overwhelmingly supports the claims of the Holy Bible that a global flood did occur. The evolutionists claim that earth's history was a long conveyer belt of "uniform" changes that took place over billions of years and dismiss any proposal of a "catastrophic" intervention by a Creator God. The debate between "catastrophism" and "uniformism" is the basic division between the two models. Is there empirical proof that a catastrophic event, such as a global flood, ever occurred? Of course there is, but have you ever read anything about it in any of the textbooks of today's liberal, humanistic, educational system?

Discoveries have been made around the world, not just at lower elevations, but on mountaintops, of great deposits of fossils of ocean creatures. These deposits are thousands of feet above sea level, found in the mountain ranges of Russia, Europe, and North America. Oh yes, even in the mountains of Montana, right along with the bones of the dinosaurs are ocean creatures, a thousand miles from the nearest ocean. Amazingly, this information didn't find its way into the textbooks. When confronted with the data from these finds, evolution scientists say these deposits are the result of an ice age, which opens up another dilemma for the evolution scientists. Most scientists have major doubts that ice ages ever occurred, because it is a concept proposed to explain the "gouging out" of the Great Lakes and mysterious striations of the earth's crust. Let's apply a little basic logic here. The polar glaciers today are only a few hundred feet high, whereas, it would require glaciers to be tens of thousands of feet high to reach the mountaintops in moderate temperature zones. A global catastrophic flood and its fossil record at mountain top levels is difficult for evolution scientists to deal with, but it must be totally devastating to them to explain away the same ocean silt and fossil record that was discovered below the desert floor in the Middle East.

In 1923, Sir Leonard Wooley, an English archaeologist, excavated the ancient cities of Sumer and Ur. He discovered the royal tombs of Ur, circa BC2400. These tombs contained 74 royal bodies, votive statues, and jewelry. This was a great discovery that confirmed the existence of cities mentioned in Genesis chapter 10, but this was not his only discovery. I think Wooley's greatest discovery was while in the search for these cities, after removing about ten feet of drift sand, he began to find silt and ocean creature fossils. He continued to dig down another 53 feet to bedrock and the fossil record continued all the way to the bedrock. His conclusion was, that during the catastrophic flood of the Bible, the waters "gouged out" the desert down to the bedrock. Here again is empirical proof of the flood, from the mountaintops to beneath the desert floor. The Babylonian deserts are 1000 miles from any ocean, and yet the evidence is there beneath the desert floor.

Here again is irrefutable, empirical evidence of the Biblical Flood that would be impossible for anyone to challenge, but, sad to say, this is not the case. Since 1923, evolution scientists have been unsuccessfully attempting to discredit this amazing find. Recently, in the "Biblical Archaeological Review," appeared a short article, that was trying to refute the discoveries of Leonard Wooley that occurred 74 years ago. Scientists, who ignore such an enormous amount of proof, base their conclusions solely on a deep wish that the flood never occurred. It is a very difficult task trying to refute empirical proof.

THE DILEMMA OF THE OCEANS

While we are discussing the flood and all that water, let's continue with the water theme. The oceans have their own story to tell. Oceanographers, in the camp of the evolutionists, are absolutely frustrated by their own theories and the mathematics that must support these theories. Let's examine the theory to its conclusion.

The popular theory today, from the viewpoint of the evolutionist, is even though the world is 4.5 billion years old, the oceans and the continents were not settled into today's approximate shape and size until about 1 billion years ago. This is because they believe that there was only one landmass and one ocean, until the landmass broke up into the continental masses of today. Temporarily, we will accept this date of 1 billion years. Apparently they must believe that there was no continental erosion for the other 3.5 billion years.

The Institute for Creation Research has developed some very interesting data from their Impact Series no.8. Consider the following:

- Deep sea drilling and seismic testing of the sediment on the ocean's floor has revealed the average thickness of ocean sediment is about 0.50 mile or (2600 feet) thick.
- The ocean floor area, around the world, is approximately 140 million square miles. To find the volume of sediment in the oceans of the world, we just multiply the two together, (0.50 mile x 140 million sq. miles), and we have 70 million cubic miles of sediment on the oceans' floor.
- There are about 383 million billion tons of continental mass above sea level in the earth today.
- The continental erosion rate of the earth's sediment and rock is about 27.5 billion tons per year that is eroded into the oceans.

If the rate is "uniform" as evolution demands, then we can easily calculate the amount the time it would take for the oceans to claim the entire global continental land mass. The calculation would be:

$$\frac{\text{383 million billion tons}}{\text{27.5 billion tons per year}} = \text{14 million years}$$

The entire continental landmasses above the ocean's level would be completely eroded away in 14 million years, and the earth would be a huge ball of water in space.

If the continents have been eroding for 1 billion years, as the evolutionists claim, then all of the land masses in the world would be eroded completely away 70 times over, and the sediment deposits of the oceans would be an amazing 100,000 feet thick, or over three times as high as Mt. Everest. I truly believe in miracles, and this one has to be the biggest one ever. The continental landmasses have NEVER once been eroded away. Aren't you glad of that?

As we can see, the problems with the evolution model of the age of the oceans are staggering, but the creation model is consistent with the evidence. According to the creation model, the ocean sediments and debris were churned and strewn over the ocean floor at the time of the worldwide flood of Noah. The most recent deposits are layered on top of the debris. For centuries after the flood, the erosion rate was significantly accelerated from what the rate is today, because the landmasses were soggy and unstable from the torrential pounding of the floodwaters. After this, the erosion rate stabilized to the present rate. It is reasonable to conclude that the oceans are very young in age, about 10,000 years or less. When you establish the criteria, and apply the preceding parameters, the creation model fits perfectly. Would it be too unkind to say that when it comes to Oceanography, the evolutionists are over their heads in very deep water?

THE UNIVERSAL FLOOD STORIES

The story of a catastrophic global flood is not only recorded in the Holy Bible, but is part of the lore of nearly every civilization around the world. The historical accounts of the Chinese version of this event are even reflected in one of their written language characters. This character pictures a boat with eight people in it, just as the Bible describes Noah and his family. The Chinese also have a written character depicting Adam and Eve in a garden, and these are both used in their written language today.

There have been 272 records around the world, in every country, depicting the ancient account of a universal flood. Many of the ancient languages, such as: Babylonian, Egyptian, Sanskrit, Chinese, Persian, Aztec, and Incan tell us of the great flood. Ancient cuneiform tablets relate the flood story in Akkadian, Sumerian, and Babylonian, as well as the sacred books of the Koran and the Holy Bible. These historical records are quite convincing about such a catastrophic world event.

Early in the twentieth century, in the deserts of present-day Iraq, a great discovery of cuneiform tablets was found. The tablets date back to the 7th century BC, during the time of the exile of Israel to the Babylonian Empire. One of these cuneiform tablets was a tablet of poetry containing 200 lines of Babylonian lore entitled *"The Epic of Gilgamesh."* This Babylonian epic tells of a worldwide flood and of an ark of safety that was built for animals and one family.

The names of the family saved in the boat may vary from civilization to civilization, and each story has its own local elements and color added in, but the basic concept is constant around the world. For dozens of ancient cultures to arrive at the same concept of a worldwide flood and of a boat full of animals, along with one family, that carried them all to safety is an astonishing phenomenon. This is a part of the historical and cultural lore of so many nations. The humanists and the evolutionists are hard pressed to even attempt to explain it all away. It is an enormous amount of empirical evidence to overcome. Perhaps the worldwide flood issue is a little embarrassing to the evolutionists, so why don't we examine their favorite topic of Anthropology in the never-ending quest for the "missing link."

ANTHROPOLOGY: MAN OR APE

In the last century, the greatest thrust of the proponents of evolution to validate their belief system has been in the field of Anthropology. This field of science seems to have been invented not for the study of man, but rather for firmly establishing man as part of the animal kingdom. I must loudly proclaim my resentment of their attempt to place mankind in the category of animals. My origins are not in the animal kingdom, but in the Kingdom of God. It is insulting to the dignity of man to be bombarded each night on some PBS telecast or a National Geographic Special about man's rise from slime. If it wasn't so serious, it would be laughable, watching people dressed up in monkey suits cavorting about in a field or in a cave pretending to be some supposed ancient ancestor of man. With all do respect to Dr. Louis Leakey, he and his followers have dedicated their lives to this one end. Perhaps some may marvel at such dedication, but I count it as a waste of precious talent and funds. The evolution scientists will go to any length to prove their belief system, fabricate any scenario, or create any plausible conclusion, in order to promote the cause.

When I was a lad, going to school was a joy and learning new things excited my imagination. I was amazed in learning all about dinosaurs and "cavemen." I listened intently as my teacher told my class all about the

Piltdown Man. The Piltdown Man was an ancient man who was a hunter of animals and a gatherer of nuts and berries. Our textbooks showed an artist's drawings of what he looked like as well as his mate and his children. The pictures of the Piltdown Man as he hunted some huge, fierce-looking animal excited my imagination. It was heartwarming to see his family gathered around their campfire, safe in their cave for the night: science at its best, teaching young children, and filling their minds with wonderful information and ideas. To fire the imagination of youngsters with mystery, awe, and truth is a marvelous achievement.

Years later, very quietly, the history of the Piltdown Man was expunged from the textbooks. It appears that the evolution scientists lied to us and committed fraud. They were so anxious to promote their theories and control what is taught in the educational systems, that they forgot things like ethics and sound scientific proof. Why was the Piltdown Man removed from scientific books of knowledge? Why isn't the Piltdown Man ever discussed any more?

The mysterious Piltdown Man was a product of a hoax and a fraud from the beginning. I remember the story that was told about a farmer in Sussex, England who buried a tooth from a pig on his farm and then dug it up again the following year. He then called the newspapers and told them of this find of a strange bone on his farm. The newspapers called in the scientists, who were all excited about this new discovery and rushed the tooth back to the laboratories for examination and testing. They tested the tooth with the Carbon-14 dating process and said that it was a tooth from an ancient progenitor of man who lived about 100,000 years ago.

From this one single tooth, the evolution scientists developed an entire scenario of what this new "missing link" looked like; his hunting skills complete with the weapons he used, and his social structure. Life-sized figures were made and put on display for the public showings along with the appropriate scenarios of the family and communal life of the Piltdown Man. This new breakthrough in Anthropology took the world by storm. The liberal humanists who control our educational system quickly embraced this new find and promptly placed the Piltdown Man and his history into the science books. So zealous were they to promote their evolution theories, that they propagated this hoax upon the entire world. This entire fraud would have continued even today if something radical and unpredictable had not happened. After about a decade of deceit, the English farmer confessed his prank and told the world of his "planting" of the pig tooth from one of his sows. The Piltdown Man fraud was very embarrassing to the evolutionists, mostly because they were caught in their scam. Why can't trained scientists tell the difference between a pig tooth that is a few years old from that of a prehistoric man? How many frauds and deceptions have been perpetrated upon the unsuspecting public?

The famous "Neanderthal Man" and all the so-called progenitors of mankind have recently undergone skeletal DNA examinations. The DNA tests showed that the coding of the "Neanderthal Man," and all the other "prehistoric men" older than Cro-Magnon, had absolutely NO link or similarity to the DNA coding of human beings. Will these so-called "human ancestors" now disappear from the textbooks of our educational system as well?

I always thought that science was based upon empirical data, tested proofs, and honest search for truth in whatever discipline any scientist endeavored to undertake. I firmly believe that most scientists are truly dedicated and honest in their science, but be aware of the others. Science is supposed to be pure, untainted, and free from any personal or political agenda. Perhaps, those who reject God the most are really those who must desperately cling to any concept, any philosophy, that gives them a feeling of belonging and purpose.

Clinging to a flawed and failed concept that is without hope is truly sad. Embracing a loving Creator God, who has exhaustively revealed Himself in His Holy Bible, would be more fulfilling. The creation of the universe and all the living creatures upon this earth are all part of God's great plan, and you are part of His plan. We have discussed some of the theories that man has devised for his own existence. Let's now examine man's theories of what God said about creation. Man even has theories on God's account of *Creation.* Let's discover God's plan together from the beginning of all things—God.

Chapter 2

In the Beginning…God

"In the beginning God created the heaven and the earth. And the earth was without form, and void; and darkness was upon the face of the deep. And the Spirit of God moved upon the face of the waters."

(Genesis 1:1&2)

The thought of an all powerful, loving, Creator God is difficult, or perhaps a little frightening to many people. We know that there are many people who believe there is a Creator God and for reasons of their own have rejected Him and pursue life according to their own dictates. There are many people who just go through life oblivious to any thought of God or His plan. A great many people, as we have seen, have sought an alternate method of creation of the universe, and of the life upon this planet. Many have chosen to believe in a heavily flawed set of theories that have been around for only a very short time. As we have seen, evolution is just another theory that never works. The options are limited: either the universe, including the earth with all of its' abundant life forms, created itself or the Creator God created all things according to His pleasure and purpose.

In our Judeo-Christian society, most people believe in God, but I might add, with varying degrees of belief, from the totally committed fundamentalists to the casual believer. America, blessed with its Constitutional rights of religious freedom, has created a nation with a multitude of religions and cults. They range from the basic main line denominations to the most absolutely bizarre cults. From such a diverse religious culture, come a variety of theories of how God created the universe. Let's explore some of these theories.

I THE RHYTHM OR PERPETUAL THEORY

We have already examined, at length, what is commonly referred to as the "Big Bang Theory." This is the favorite theory of the evolutionists: a self-creating universe without the acknowledgment of a Creator God. This is a process that even denies and rejects the very idea of a Supreme Being.

The "Rhythm or Perpetual Theory" doesn't care if the universe was exploded into existence by a Creator God or was self-created. This is sort of a hybrid theory, a transition between Evolution and Creation. This theory is more concerned about the physical action of the universe than of its' origin, although it does tend to drift heavily toward the side of evolution. The "Perpetual Theory" takes the "Big Bang Theory" a step further.

Once the universe was set in motion by God speaking it into existence or from a gigantic exploding ball of hydrogen gas from which all things evolved, this theory claims that the universe will expand out to a certain limit and then collapse in on itself, and then start all over again, perpetually expanding and contracting.

Let's use a children's' toy as an example. Do you remember playing with a "paddle ball" when you were a child? It was a small wooden paddle that had an elastic cord attached to the paddle at one end, and a little rubber ball attached to the other end of the cord. You were supposed to hit the ball and it would shoot out until the elasticity of the cord would finally overcome the momentum of the ball, and the ball would come sailing back, only to be hit again and again.

Scientists, who adhere to the "Perpetual Theory" claim that the universe is expanding in every direction, like an exploding fireworks display. They claim the expansion of the universe is "slowing down" and some day will come to a complete stop, then will begin to contract upon itself until the universe will implode in on itself to create another "big bang." This is a universal, perpetual-motion machine when the universe expands and contracts upon itself, creating an infinite big bang process that is never-ending. Here is another theory that can never be proven. This probably is a computer-driven model designed by some with entirely too much time on their hands. Perhaps, it would take tens or even hundreds of billions of years to happen. If other scientists disagree with this scenario, then they would add another hundred billion years to the program and then it should work just fine.

We should not worry too much about this particular theory because the universe has never collapsed in upon itself and we are still all here. This theory does not deal so much with the creation of the universe, as it does with what happens to it once it was created. This theory, like most theories, is based on wishful thinking and little substance. There is no hope of any of the parameters of this theory ever to be in place, but it does contain enough plausible sounding logic to springboard a host of new theories. The God of this universe has been relegated to many strange roles. Let us explore another of the theories of the creation of the universe.

II THEISTIC EVOLUTION

Because of the effects of the theory of evolution upon the Christian community over the years, some liberal theologians have adopted some of the theories of evolution and wedged them into established fundamental beliefs here and there, along with a few of their own pet theories, and have created some new "hybrid concepts."

One of these concepts is called the "Theistic Evolution Theory." Its basis is simply a merger of the Creation and Evolution models. The basic premise is that God created the universe as the Bible states. God spoke the universe into existence, and setting all things in motion. The billions of stars were flung into space, forming millions of galaxies and nebulae. Our own galaxy, the Milky Way, and our solar system were formed at His pleasure. The only known planetary system in the universe was created, which includes the earth. In 1992, astronomers found a few planetoid bodies orbiting around "pulsar stars." Astronomers have never found any other planetary systems like ours in any of thousands of galaxies catalogued.

The advocates of "theistic evolution" supposedly believe in God and are trained in God's Word, but amazingly adopt all the tenets of evolution concerning how life began on this earth. They must believe that the Creator God was too tired to create anything else after creating the universe, or perhaps He just didn't know how to create life. This is Theology at its liberal best. It is very sad to see, perhaps, very sincere men and women deceived by the "great lie" of evolution that is so widely promoted in our time.

III CREATION (FATALISM)

Another variation of this theme is that God created the heavens and the earth and all of the myriad life forms on this planet just as it are recorded in the Bible. God set all the mechanics into motion for the universe, including the earth, and then He turned it all loose. He no longer is involved in the affairs of the life forms on this planet. Of course, this is contrary to what God says in His Holy Scriptures.

This is the "Fatalistic" approach: that God created all things and abandoned His creations to an uncertain fate. This theory speaks of a creation without a plan or purpose for existence, and of an impersonal, uncaring God who created all things as a whim. The "fatalism" expressed here is a failure to see God's hand on the lives of men and women everywhere, and a failure to believe and trust in God's Word. This theory of "what will be, will be" is based on a personal failure to reach out to a personal, loving God who most certainly wants to be involved in each of our lives.

This theory, by its very nature, is contradicting itself. It proclaims a living Creator God who created all things, the universe, and the earth with its vast array of life. Then, it proclaims the Godless theory of evolution took over to establish life, as we know it. This is a very strange combination of concepts, which tells me that those who believe in this theory really do not have a commitment to either a Creator God or to Evolution. Defending one's position with this theory would be difficult. One would get criticism from both sides, and acceptance from none.

This theory, from a theological standpoint, negates any validity of God's Holy Word, and declares the birth, the life, the death, and the resurrection of Jesus Christ to be irrelevant. It strips away the basic, fundamental truths of God's Word, and discards God's own concise description of His creation in favor of a flawed theory of man.

From the secular standpoint, this is a very risky position, because one must admit that there is a Creator God, and that He did create the heavens and the earth. This is quite an admission for one who believes in evolution, or at least partially believes in it. This is a very high, wobbly fence that believers of this theory sit upon. They seem to want the best of both worlds, but cannot put their complete trust in either side.

IV THE LITERAL OR PURIST THEORY

The "Purist Theory" is the bedrock belief of fundamental, Bible-believing Christians and Jews. These are people who believe in the "literal" interpretation of the Holy Scriptures, as God had commanded them to be written. To the believer, God's Word is not a theory, but The Truth for the ages. The Holy Bible is the revealed Word from the God of the universe. It is a "love letter" from a loving Creator who exhaustively reveals Himself to all who seek Him. I need not try to define who or what God is. To do so, I would have to be as God.

As the Bible opens with, *"In the beginning God created the heaven and the earth,"* God does not explain who He is, but tells us what He did. There are no philosophic arguments by God to explain His existence. The Creator gives us information that we are able to comprehend: a beginning. We are able to understand infinity, because it always has a starting point. For example, if we were standing on the seashore looking toward the horizon at a beautiful sunset, our line of vision would begin from our location to a point tangent to the earth, then extend into outer space or infinity. This is called infinity and we are able to understand it, but the comprehension of eternity is beyond human understanding. Our Creator gives us a "beginning" to start with, so that we may grasp His revelation. Since we humans are finite beings, we are limited to the realm of the space/time continuum. But in the Bible, God reaches out to us with eternal concepts, and extends eternal hope to everyone.

The "Purist" believes exactly as the Bible states, that God created the heavens and the earth about BC4004, with all of its life forms, which were created in six days; and God rested on the seventh day and made it holy.

> *"Thus the heavens and the earth were finished, and all the host of them.*
>
> *And on the seventh day God ended his work which he had made; and he rested on the seventh day from all his work which he had made."*
>
> (*Genesis 2:1&2*)

The "literal" interpretation of the Scriptures is quite comforting to believers, because they are putting their total trust in God's Word. If you trust that the God of the universe knows exactly what He is talking about, then why should there be any doubt? They use the old familiar phrase "God means what He says, and says what He means." I believe that the old adage about faith would apply here.

"For those who do not believe in God, no amount of explanation is possible.

For those who do believe in God, no amount of explanation is necessary."

Why do so many people doubt the Word of God, and put their trust in the theories of men? It is amazing to think that some mere mortal would know more than the Lord God Almighty, creator of all things, including us mortals.

> *"God is not a man, that he should lie; neither the son of man, that he should repent."*
>
> *(Numbers 23:19a)*

After all, who else was there when God created the universe, and who else could give an honest report on the greatest event of all time: creation? The creation of this earth, and its vast array of life, is so astounding that it takes ones breath away just to consider the enormous problems involved in such an undertaking. This creation is a grand design, lovingly created by God who has a wonderful plan for everyone who will seek Him.

There are some people today in the New Age cults who actually claim that they are gods. What arrogance to make such claims. Where were they when the universe was created? Why was the earth created with its myriad life forms? Why is man here, and how does he fit into God's plan? God states flatly that it is His creation, His plan, and for His purpose. God spoke to Job out of a whirlwind stating these very same thoughts in the Bible.

> *"Where wast thou when I laid the foundations of the earth? Declare, if thou hast understanding.*
>
> *Who hast laid the measures of it, if thou knowest? Or who hath stretched the line upon it?*
>
> *Whereupon are its foundations fastened? Or who laid its cornerstone.*
>
> *When the morning stars sang together, and all the sons of God shouted for joy?*
>
> *(Job 38:4-7)*

God establishes in His Word all the facts we need to believe and trust that everything He said is true, which will stand the test of time. Throughout the ages, there have been scientists, philosophers, and writers who have diligently tried to disprove God. Even though all have failed, they still keep trying. There always has been and always will be a never-ending supply of dissenters.

V EX NIHILO

The phrase *"ex nihilo"* is a phrase that perfectly describes the Purists' Theory. It comes from two Latin words, *"ex"* which means "out of" or "from," and the second word is *"nihilo"* which means "nothing" or "nothingness." When used in the context of creation, its meaning is quite eloquent. When God created the heavens and the earth, He created them "out of nothing" or "from nothingness." God created all things by speaking them into existence, according to His plan, and for His pleasure.

> *"Thou art worthy, O Lord, to receive glory and honor and power; for thou hast created all things, and for thy pleasure they are and were created."*
>
> *(Revelation 4:11)*

In AD1650, James Ussher, who was an Archbishop in the Protestant Church of Ireland, published his study on the Bible. This exhaustive work was called *"The Chronology of the Old and New Testament,"* which was written in Latin. In his research, he calculated the genealogies and lineages in the Scriptures to establish the time frame in which Adam was created. He concluded that God created Adam in the year BC4004, exactly four thousand years before the birth of Jesus Christ.

John Lightfoot, a Hebrew scholar and Vice Chancellor of Cambridge University, only a few years after Archbishop Usshers' publication, redid his calculations. Lightfoot's findings were that the creation of the earth occurred at 9 o'clock on Sunday morning of October 23 in the year of BC4004.

Eusebius, the Bishop of Caesarea in the fourth century, was considered to be the father of church history. Eusebius was not only a prolific writer, but also made enormous contributions toward the establishment of an acceptable canon of the Scriptures. Under the direction of the Roman Emperor Constantine, Eusebius organized and classified many aspects of church doctrine and chronology. One of his most interesting calculations was that the creation of man occurred about the year BC4000.

Martin Luther was a Roman Catholic priest and theologian in Germany. Luther was best remembered for posting his "Ninety-five Theses" on the door of the All Saints Church in Wittenberg, Germany, in AD1517. He, too, established the creation at about BC4000, independently from the Roman Catholic studies, which he considered to be too corrupt to be trusted.

Pope Gregory assigned the task of devising a new calendar to a Roman monk by the name of Dionysius Exiguus, circa AD6th century. Dionysius calculated the birth of Jesus to be in the Roman year 754, and called it "Anno Domini," or AD1, which in Latin means "the year of our Lord." Scholars, over the centuries, have maintained that his findings were in error, because he miscalculated the time of the reign of Julius Caesar. Historians wisely have not readjusted the calendar years to relocate the date BC/AD, but have reset the date at BC4, for the birth of Jesus, instead of AD1. This would cause worldwide chaos and confusion, if the year AD1996 suddenly became the year AD2000.

The window of error found in the Gregorian calendar was about seven years. So where do you place the birth of Jesus? The range was between BC7 and BC/AD to place the birth of Jesus. The most generally accepted date was set at BC4, but as expected, scholars and historians have disputed this finding. My own research has found that BC3 was the birth of Jesus. Can you imagine the chaos, every time some historian found a piece of evidence that would change the calendar year? It was wise not to change the calendar year.

Scholars finally set a generally accepted date of BC4, as the birth of Jesus. Therefore, the four thousand years between the creation of Adam to Jesus' birth was now reset to Adams' creation in BC4004, and the birth of Jesus at BC4.

From the viewpoint of the "Purist Theory," the creation of the heavens and the earth occurred in BC4004. Then God created all the life forms upon this earth in the following six days, ending with His crowning achievement, the creation of Adam. They strictly adhere to the time frame of each creative day, as being a twenty-four hour day. After creating Adam from the dust of the earth, God rested on the seventh day.

VI THE GAP THEORY

The "Gap Theory" is identical to the "Purist Theory," except for one major point. The major difference is when God created the heaven and the earth, it was millennia or perhaps eons before God recreated the earth. The theory states there is a "time gap" of an unknown time period between creation of the universe in Genesis 1:1, and the recreation of the earth in Genesis 1:2.

The scenario for the “Gap Theory” is that a time gap exists, of an indeterminate amount of time, between the actual creation of the universe, and the “restoration” of the earth. The beginning of all things was eons ago. The universe was complete, including the earth, and then God did a special recreation on the earth, about 6000 years ago. He created life upon this planet, infinite in variety and wonderfully made. I personally embrace the “Gap Theory,” not because it seems to be the prudent choice, but after thirty years of Bible study and fifteen years of teaching the Bible, I have arrived at this conclusion based solely on the basis of Scriptural revelation. I would like to share with you the Scriptural reasoning behind my choice of the “Gap Theory.”

SIN ENTERS THE UNIVERSE

Let your imagination drift backward in time, even before the beginning of time. There are no stars; there is nothing in the universe except God and His created angels. Then something wonderful happened: God spoke; and a universe was filled with brilliant, shining stars, as the angels looked on with awe, and trembling.

> *“In the beginning God created the heaven and the earth.”*
>
> *(Genesis 1:1)*

God told the angels about His great plan of creation, particularly about this one, beautiful, water planetoid called Earth. God told them that this planet was to be inhabited by beings, perhaps by the angels themselves, and the great angelic host shouted for joy.

> *"Whereupon are its foundations fastened? Or who laid its cornerstone.*
>
> *When the morning stars sang together, and all the sons of God (angels) shouted for joy"*
>
> *(Job 38:6&7)*

The angels rejoiced in the presence of God, giving Him praise for the awesome beauty that He had created. But there was one archangel who rebelled against God. His name, *"Helel,"* in the original Hebrew, was called *"Shining One,"* or *"Day Star,"* but we know him as Lucifer. The name Lucifer is not found in the original Hebrew text, but was introduced by St. Jerome, who translated the Hebrew and Greek Bible into Latin, called the Vulgate Bible, in about AD405, at the Church of the Nativity in Bethlehem.

LUCIFER FALLS

Lucifer, being the most beautiful of all of God's creations, was full of pride and aspired to be as God. Sin was found in Lucifer, and he caused a rebellion in heaven, and persuaded one third of the angels to follow him. The angels true to God overcame them, but those who followed Lucifer were judged and cast out of heaven.

> *"How art thou fallen from heaven, O Lucifer, son of the morning! How art thou cut down to the ground, who didst weaken the nations!*
>
> *For thou hast said in thine heart, I will ascend into heaven, I will exalt my throne above the stars of God; I will sit also upon the mount of the congregation, in the sides of the north,*
>
> *I will ascend above the heights of the clouds, I will be like the Most High."*
>
> *(Isaiah 14:12-14)*

Lucifer wanted to be the God of the universe, but was cast out of heaven, down to the earth, with all of the fallen angels with him. Those angels that sinned against God were removed from His presence forever. Lucifer, now called Satan, by God's grace, was permitted to visit heaven to be the accuser of the saints and to visit the earth to be the tempter of all mankind. This was all done by God's permissive will and in accordance with God's wonderful plan for His creations.

> *"And there was war in heaven; Michael and his angels fought against the dragon, and the dragon fought and his angels,*
>
> *And prevailed not, neither was there place found any more in heaven.*
>
> *And the great dragon was cast out, that old serpent, called the Devil and Satan, who deceiveth the whole world;*

he was cast out into the earth, and his angels were cast out with him."

(Revelation 12:7-9)

When God created the heaven and the earth, every creation was pristine and perfect. God's pristine world was designed; first for sinless, angelic creatures, and secondly for that creation which was to be His crowning achievement in the entire universe: man. The earth was formed by God to be inhabited by His "companions."

"For thus saith the LORD who created the heavens, God himself who formed the earth and made it; he hath established it, he created not in vain, he formed it to be inhabited: I am the LORD, and there is none else."

(Isaiah 45:18)

Humiliated by his defeat in heaven; Satan, now on earth with his army of fallen angels, plans revenge against God. Satan knew the plan of God: to create a companion that he could fellowship with, a spirit- man. Satan knew that God was going to give man the title deed to this beautiful planet, and he and his army of fallen angels might be forced to leave. Satan's rage grew within him: pride, jealousy, anger, and revenge, churned within him like a seething caldron. A plan formulated within him; he thought that if he could not have this beautiful planet, then he would destroy its beauty, and make it so uninhabitable that God would not create His spirit-being: man. So Satan and his minions set out to destroy the earth in a vain attempt to foil God's plans to create a paradise on earth for man.

"Who made the world like a wilderness, and destroyed its cities, who opened not the house of his prisoners?"

(Isaiah 14:17)

With the earth now destroyed, we come to the end of the "time gap" between Genesis 1:1 and Genesis 1:2. This scenario now takes us to the devastated, formless, and darkened earth, where God now has to renovate or recreate the earth in the second verse. This definitely is not the condition of the planet Earth that God had originally created; but we are now given the picture of a loving God, patiently and purposefully renovating it for His companion: man.

THE RECREATION OF PLANET EARTH

In Genesis 1:2, we find this beautiful, blue planet called Earth, is now just a blackened shapeless hulk, totally covered in darkness. As the previous Scripture stated: Satan had "*made the world a wilderness*." With the fall of Satan, sin had entered the universe and now had destroyed and tainted the earth. God had originally created the earth pristine and holy in preparation for man, but it was a ruined globe that had to be renovated. The earth was now cursed with sin, and would not be redeemed from that curse until after all sin is judged at the Great White Throne of Judgment (Revelation 20:11-15). When all sin is cast into the Lake of Fire, then the earth will be redeemed and cleansed by fire (2 Peter 3:10-13), and made holy by God for the New Jerusalem (Revelation 21:1-3).

Since God did not create junk, He began to renovate the earth all in accordance with His plan and all of these events are in the foreknowledge of God. As we continue the scenario, we are told what the next step was in God's plan in dealing with this attempt by Satan to thwart the grand plan for mankind.

> *"And the earth was without form, and void; and darkness was upon the face of the deep. And the Spirit of God moved upon the face of the waters."*
>
> *(Genesis 1:2)*

As we can see here, the earth was restored, renewed, renovated, recreated, or whatever term seems appropriate. The earth had to be restored. God's six days of recreation is the account given in Genesis. After the six days of recreation (which we will cover in later chapters), God then "planted a garden" eastward in Eden. The whole earth was cursed by sin because of Satan's destruction, but God created a sinless paradise for man in the midst of the corrupted earth. The garden of God was called Eden and four rivers compassed it about. The four rivers are called; first the Pishon, second; the Gi'hon, third; the Hid'dekel, and fourth; the Euphrates (Genesis 2:10-14).

Within the four boundary rivers was sinless perfection, but outside those boundaries was sin and death. God had created a microcosm of heaven on earth for His greatest creation. Adam was "placed in the garden" by God

> *"And the LORD God took the man, and put him into the garden of Eden to till it and to keep it."*
>
> *(Genesis 2:15)*

The very first act of God's mercy and judgment can be found right here in the garden of Eden. Satan and the fallen angels were not destroyed for their actions in heaven, or upon the earth, but were shown mercy by God. If God had destroyed them, do you think the faithful angels would now serve God out of love or because of fear? God's judgment was also present here, because He now has given man the ownership of the planet. God also commanded Adam and Eve to replenish the earth in lieu of the former angelic inhabitants. There had to be previous occupants of the earth for God to command the first two human beings to REPLENISH the earth. How long this previous occupation was, no one knows.

> *"And God blessed them, and God said unto them, Be fruitful, and multiply, and replenish the earth, and subdue it; and have dominion over the fish of the sea, and over the foul of the air, and every living thing that moveth upon the earth."*
>
> *(Genesis 1:28)*

This concludes my scenario for validating the "Gap Theory" as a viable theory based on a Scriptural premise. Perhaps this explanation may be an illumination for you, something that you never before considered, or maybe it only confirms your own conclusions. In any case, it has been an enjoyable time exploring "my scenario" of the "Gap Theory," traveling together with you, on a journey in the search of truth.

VII THE SIX-DAY THEORY

There is one final theory that I must share with you. There are many avenues that one may pursue in the study of Biblical prophecy. Each study is extremely fascinating within itself, but none is more helpful in giving an overall grasp on God's plan and purpose, than the "Six-Day Theory."

There are many exciting studies such as; Eschatology, Numerics of the Bible, Gematria, or the latest astonishing revelation called "Equidistant Letter Sequencing." These are all marvelous disciplines of Bible study, but this is not the format to delve into them with you.

The Six-Day Theory is a theory that is directly connected to the creation. Each day of the six days of creation represents one thousand years of man's history. Man has toiled and struggled now for 6000 years, from BC4000 to AD2000. As God rested on the seventh day, man will rest for 1000 years in the next Dispensation, which is called the Millennium. Dr. Jack Van Impe, the noted Bible scholar and prophecy teacher, has researched the origin of this ancient theory. He found it to be a very old theory that has been used throughout history to establish where in time certain Biblical events would occur. There have been many scholars throughout the centuries, both Jews and Gentiles, who have used the Six-Day Theory as a method to calculate the coming of the Messiah. The following is listing of some of these scholars who have used this theory:

- In the 4th century BC, a book called "The Book of Enos" was written. The writer stated that the expected Messiah would come in 2400 years from his time period.
- In the 2nd century BC, rabbinical scholars had calculated that the Messiah would come in 2200 years.
- In the 1st century AD, a book called "The Epistle of Bartholomew," written by the Apostle, predicted that Jesus, the Messiah, would return 2000 years after his birth.
- In the 1st century AD, a book called "The Epistle of Barnabas," written by the traveling companion of Paul, stated that Jesus, the Messiah would return 2000 years after his birth.
- In the 4th century AD, Eusebius, the church historian, calculated that the Messiah would return in about 1700 years.
- In the 17th century, Martin Luther, the great reformer, had also calculated that the Messiah would return in the year AD2000.

These scholars, and many more, used the Six-Day Theory to calculate that Jesus, the Messiah, would return in the year of AD2000. The short version is: from Adam to Jesus was four days, or 4000 years; and from the birth of Jesus to his Second Coming will be two days, or 2000 years. After a short seven-year period called the "Tribulation Period," Jesus will rule the world in peace and harmony for 1000 years, in a period called the "Millennium," the seventh day of rest.

THE DISPENSATIONAL PLAN OF GOD

It would be prudent at this time to inject the Dispensations of God's plan to help clarify the terminology. Each phase of God's plan is called a Dispensation or an Economy. Each Dispensation represents a particular time frame in the history of man. Let's review these time periods as follows:

1st Dispensation—"The Age of Innocence," or "Creation"

2nd Dispensation—"The Age of Conscience"

3rd Dispensation—"The Age of the Patriarchs"

4th Dispensation—"The Age of the Law"

5th Dispensation—"The Age of Grace"

6th Dispensation—"The Tribulation" (7 years)

7th Dispensation—"The Millennium" (1000 years)

God's perfect number is seven, and it is not surprising that God's plan for man has seven stages. Every stage in man's history is proclaimed in the Word of God. We presently live at the very end of "The Age of Grace." The unbelieving world will soon suffer the wrath of God in the seven-year "Tribulation" period, followed by Jesus ruling the world for one thousand years: the period called the "Millennium."

We might consider an 8th Dispensation which could be called "Eternity with God." The number eight in Bible numerology means "New Beginnings." After the final judgment of all sin, at the Great White Throne of Judgment, God will cleanse the earth with fire, and the New Jerusalem will be brought from heaven and placed on a holy earth, where

God and His redeemed children will live together forever and ever. It is the "new beginning" where truly *"the old things are passed away, and all things will become new."*

A DAY IS AS A THOUSAND YEARS

The ancient scholars who searched the Scriptures daily, keyed in on one verse to formulate their theories. In the Psalms, this verse gave them the clue to use a thousand years for each day of creation. They found that many of the confusing prophecies now made perfect sense. This revelation opened up new avenues of research. The verse that I am speaking of is as follows:

> *"For a thousand years in thy sight are but as yesterday when it is past, and as a watch in the night."*
>
> *(Psalms 90:4)*

Each day represented a thousand years, so if one could calculate the time when Adam was created, add six thousand years to that date. You would then know the date of the end of this phase of God's plan. When the New Testament was canonized, scholars found confirmation of their theory in the New Testament. The so-called Six-Day Theory became widely used. As we have seen earlier, many well-known scholars utilized this means of calculating prophetic events. The final judgment of all things at the end of the 7th Dispensation, is undeniably connected to the Six-Day Theory. The Great White Throne Judgment is the conclusion of the "Seventh Day," when all the sin of the universe is judged. Let's examine the verses together that speak of judgment and days.

> *"But the heavens and the earth which are now, by the same word are kept in store, reserved unto fire against the day of judgment and perdition of ungodly men.*
>
> *But, beloved, be not ignorant of this one thing, that one day is with the Lord as a thousand years, and a thousand years as one day."*
>
> *(2 Peter 3:7&8)*

These verses make it quite clear, that in order not to be ignorant of God's Word, we must consider using this concept. To me, this is not just a theory, but a command from God to use it.

VIII AFTER SIX DAYS…

There are many examples of the Six-Day Theory in the Holy Scriptures, some are in the Old Testament and some are in the New Testament. Some of these stories are quite subtle, while others are very easy to understand, but they are all prophetic in nature.

Let's consider some examples of the "six days." First, let's go to Mount Sinai where Moses is receiving the Ten Commandments. Here God forever establishes the six days of work and one day of rest concept. It is patterned after God's days of creation.

> *"Remember the sabbath day, to keep it holy.*
>
> *Six days shalt thou labor and do all thy work;*

But the seventh day is the sabbath of the LORD thy God; in it thou shalt not do any work...

For in six days the LORD made heaven and earth, the sea, and all that in them is, and rested the seventh day; wherefore, the LORD blessed the sabbath day, and hallowed it."

(Exodus 20:8-11)

There is no calculation involved here, but it is one of the first applications of God's Word connected to His creation of the earth.

NA'AMAN THE LEPER

The next example is very subtle, but still has the "end time" prophetic flavor to it. In 2 Kings 5:1-14, is the story of Na'aman, the captain of the armies of Syria, who sought out Elisha, the prophet of God in Israel. Na'aman was a leper, and had heard that through this holy man, many people had been healed. He wanted an audience with Elisha, but the prophet refused to see Na'aman. Elisha sent Na'aman a message which said *"go wash in the Jordan seven times, and thy flesh shall come again to thee, and thou shalt be clean" (vs. 7)*. Now let's interpret this story in light of the prophetic "Six-Day Theory."

Na'aman symbolizes the gentile world, that has been given the message from a man from God, to be washed or cleansed in the Jordan. Each of us must pass over from this world through the Jordan (death) into the "promised land," or heaven. We shall be cleansed after "six times" or six thousand years, and on the seventh time we shall be cleansed of our leprosy (sin), made pure and whole by the power of God. This is an Old Testament picture of the cleansing power of the blood of Jesus Christ, and the promise of being made whole and pure after six thousand years, or after "six days."

THE MOUNT OF TRANSFIGURATION

There are several examples in the Old Testament that we could explore, but now I would like to go to the New Testament for our next example. The story of the Transfiguration on Mount Tabor is the perfect example that is very straightforward in its concept.

> *"And after six days Jesus taketh Peter, James, and John, his brother, and bringeth them up into an high mountain privately,*
>
> *And was transfigured before them; and his face did shine like the sun, and his raiment was as white as the light.*
>
> *And, behold, there appeared unto them Moses and Elijah talking with him."*
>
> *(Matthew 17:1-3)*

The narration speaks for itself as to what happened, but why did it happen? What was the purpose of this event, or what message does it bring? This was an astonishing miracle that was beheld by the inner circle of the Apostles. They were charged not to reveal what they had seen, but we are privileged to imagine it through the Scriptures. This is a perfect picture for us to interpret through the Six-Day Theory.

The message is first of all given to the three Apostles, who are symbolic of the church. Notice that the verses begin with the term *"after six days,"* or after six thousand years. The scene that they witnessed symbolized what would occur in God's plan after six thousand years from Adam. Jesus was transfigured before their eyes: from his earthly body to His glorified heavenly body. Along with Jesus were the glorified figures of Moses and Elijah. The glorified bodies of Moses, Elijah, and Jesus are symbolic of all believers, since Adam, who have trusted in God and in the resurrection power of Jesus.

The promise of eternal life is the message throughout the Scriptures that burns in the heart of every believer. This will occur when Jesus returns for His Bride—the Redeemed—at His second coming. Our mortal bodies shall be changed into glorified bodies—in the twinkling of an eye; from the earthly to the heavenly; from corrupt to incorruptible; from mortality to immortality. This hope of eternal life is beautifully pictured here, and promised in the following verses (1 Thessalonians 4:13-18) and (1 Corinthians 15:51-57).

The glorified figure of Elijah represents the Prophets; that of Moses represents the Law; and Jesus represents the Church. These three figures span the ages of man; from Adam to the Second Coming of Jesus Christ which is called the Rapture. Everyone that belongs to God will be taken out of this earth, both the living and the dead. All those who have rejected God, and the plan of salvation through His son Jesus Christ, will be left behind to face God's wrath and judgment.

To summarize this scenario of the transfiguration, we will translate it into the prophetic meaning given to us by the application of the Six-Day Theory. The message is given to the church that after six thousand years from Adam, the Rapture will occur. At the Second Coming of Jesus, our bodies will be changed into glorified bodies and the Redeemed we shall be as He is. This will end the Church Age and usher in the dreaded Tribulation, a seven-year period of God's wrath and judgment of the unbelieving world.

THE CRUCIFIXION OF JESUS CHRIST

The Six-Day Theory is very flexible, and is not restricted to days and years, but may be used in a variety of ways. Let's investigate a different approach to glean new prophetic revelations, but still maintain the basic principals of the theory. We can expand our understanding of the richness of the Scriptures by being alerted to the usage of the number "six" in the text. In the numerics of the Bible, the number "six" is the number for "man," and the number "seven" is the number for "God's perfection." We will explore the use of "six waterpots" in the section "After Two Days," but in this example we will see how "six hours" may enrich our understanding of this very familiar story. The narration of the crucifixion of Jesus of Nazareth has been known throughout the world, in every generation, for two millennia. So let's imagine that we are spectators, as we return to this gruesome scene, at a hill called Golgotha in Jerusalem.

At the third hour (9 o'clock) in the morning, Jesus of Nazareth was being nailed to a cross as an enemy of the Roman Empire. Two thieves were also crucified, one on either side of Jesus. As the day wore on, about the sixth hour (noon), darkness covered all the land (Matthew 27:45), until the ninth hour (3 o'clock). Jesus died on the ninth hour, and his lifeless body remained on the cross for another hour, until his friends were able to receive permission to remove him from the cross. It was truly a cruel and bloody scene. As we observe these events, we are reminded that our quest is not to describe the scene, but to interpret the prophetic meaning behind the picture before us. Let's explore this tragic scene in the light of our prophetic formula of the Six-Day Theory and, perhaps, receive some new insight into this very brutal event.

Promptly at 9 o'clock in the morning, the Temple priests start to select and process the lambs for the sacrificial offering of Passover. They begin to slaughter the sacrificial lambs about 3 o'clock in the afternoon. All of the lambs that were used for the Temple sacrificial rituals were raised in Bethlehem. Jesus, the Lamb of God, was being nailed to the cross on Golgotha that same hour. Sin and death, and sickness and disease, were all nailed to that old rugged cross of the suffering Savior.

> *"Surely he hath borne our griefs, and carried our sorrows; yet we did esteem him stricken, smitten of God, and afflicted.*
>
> *But he was wounded for our transgressions, he was bruised for our iniquities; the chastisement for our peace was upon him, and with his stripes we are healed."*
>
> *(Isaiah 53:4&5)*

Blackness covered all the land about noon, so that not even the angels of heaven were permitted to see the suffering of Jesus as he received the punishment for the sins of all mankind. For "six hours," Jesus suffered the agonies of the "six thousand years" of man. Jesus paid the penalty for the sinful existence of all mankind since Adam.

Jesus told the thief on the cross, *"this day you shall be with me in Paradise."* There were only a few more hours left in the day, but Jesus kept his promise. Jesus died on the ninth hour. His soul and Spirit went to Paradise, where he made a spectacle of Satan, and took the keys of Hell and Death. Jesus then took everyone, who trusted in God, out of Paradise to heaven, and presented them to the Father.

In our prophetic scenario, Jesus paid the "wages of sin" during the "six hours," or 6000 years. On the "seventh hour," or the 1000 years of rest, just as the lifeless body of Jesus was at rest in the "seventh hour," the Body of Christ will be at rest in the heavenly Paradise. When Jesus will come for His Bride, she will be dead to sin, secure, and at rest in His arms. Jesus will lovingly take His Bride and present her to His Father. Satan will no longer have any power over the Redeemed, because they have been changed into their eternal bodies, and are forever at rest in the presence of God. This gruesome scene of the crucifixion presents a beautiful, prophetic picture of eternal hope, for all who are trusting in what Jesus did on the cross of Calvary, for all of mankind.

IX AFTER TWO DAYS...

There is another application of the Six-Day Theory that uses the fifth and sixth days by themselves. Let's define the days once more to clarify their meaning.

Days One & Two or 2000 years	From Adam to Abraham
Days Three & Four or 2000 years	From Abraham to Jesus
Days Five & Six or 2000 years	Jesus' First & Second comings
Day Seven or 1000 years	Jesus rules the world as King of Kings

Our study now focuses in on the Fifth & Sixth days, from BC/AD to AD2000. Prophetically, these two dispensational days are known as the "last days." The Seventh day is known as the "Day of the Lord." Often, the phrase, "end times," is used in the Bible to depict the time frame at the very end of the Sixth day, or just before the year AD2000. This is our present lifetime, and it is a very exciting time in which to be living. It is within this time frame that our studies now carry us.

Prophetically, the phrase, "end times," is depicted as two years, when a "time" is equivalent to one year and the plural "times" would be two years (Daniel 7:25), (Revelation 12:14).

Consider, also, the phrase, "last days," which has a similar meaning. The prophetic reference to a "day" also means one year and the plural would be two years (Numbers 14:34), (Ezekiel 4:6). Each of these phrases, “end times,” and “last days,” refer to the last years of Day Six Dispensation, or near to AD2000.

HE SHALL RAISE US UP

The phrases often used in the Scriptures, *“after two days”* or *“on the third day,”* are used much like the phrase, *“after six days,”* but refers only to the last 2000 years. These prophetic “end time” word pictures are not restricted to the New Testament, but occur in the Old Testament as well. The Jews are looking for their Messiah in the “last days,” and, in particular, at the end of the Tribulation period. The Jewish remnant shall call upon the Lord for salvation: *“they shall look upon me whom they have pierced”* (Zechariah 12:10). Let’s explore an Old Testament prophetic example given in the Scriptures.

> *“Come, and let us return unto the LORD; for he hath torn, and he will heal us; he hath smitten, and he will bind us up.*
>
> *After two days will he revive us; in the third day he will raise us up, and we shall live in his sight.”*
>
> *(Hosea 6:1&2)*

Hosea, who lived eight centuries before the birth of Jesus, received this prophetic message from God about the remnant of the Jews in the Tribulation period. Israel will desperately cry out for the Lord to save them from annihilation, during the battle of Armageddon. These amazing verses are not only in the Old Testament, but they speak of events that are yet to happen, even twenty-eight centuries after Hosea penned his book. Verse two contains both phrases: *"after two days"* and *"on the third day,"* which is extremely rare. These verses also speak of reconciliation of the Jews to God, calling upon the Lord Jesus for salvation, and the resurrection of the Jewish believers at the Rapture, and living eternally with the Lord. Prophetically, these verses by Hosea have always fascinated me, and as you can see, are very rich in their content.

THE RESURRECTION OF LAZARUS

The next prophetic picture takes us to the familiar story of Jesus raising Lazarus from the dead. Jesus was preaching in the Galilee region, when he received word that his dear friend Lazarus, who had been sick, was now dead. Something very strange then happened; Jesus did not immediately leave to see his dear departed friend. Why did Jesus wait? What was his reason for not rushing to his side? Is there more to the story than the already wonderful narration? This is not only a story of love and compassion, but also one of prophetic assurance of our eternal life. Let's look deeper into the story, with a prophetic eye, and see the eternal promises of God, that give us hope.

> *"Therefore, his sisters sent unto him, saying, Lord, behold, he whom thou lovest is sick.*
>
> *When Jesus heard that, he said, This sickness is not unto death, but for the glory of God, that the Son of God might be glorified by it.*

Now Jesus loved Martha, and her sister, and Lazarus.

When he had heard, therefore, that he was sick, he abode two days still in the same place where he was."

(John 11:3-6)

Let's analyze this narration in the context of the prophetic concept of *"after two days."* Lazarus represents the church, a friend of Jesus, who has a sickness that is even unto death (sin). Jesus, who is in a far away place, knows of our deadly sickness, chooses to wait two days (2000 years) to heal us and raise us from the dead (Rapture). Jesus, after two days, came to Lazarus, and with tears and compassion, gave a loud shout, *"Lazarus, come forth"* (John 11:43). Lazarus was raised from the dead, and was once again in the presence of Jesus. The church will also experience victory over sin and death, when Jesus comes for His Bride with a shout (1 Thessalonians 4:13-18) and we shall all be changed.

"So, when this corruptible shall have put on incorruption, and this mortal shall have put on immortality, then shall be brought to pass the saying that is written, Death is swallowed up in victory.

O death, where is thy sting? O grave, where is thy victory."

(1 Corinthians 15:54&55)

When Jesus shouted at the tomb in the mountain, and called for Lazarus to "come forth" from the dead, in my opinion, if Jesus did not call Lazarus by name, every dead body in that mountain would have been raised from the dead. Jesus is the Lord of the resurrection, and the eternal hope of all mankind.

"Jesus saith unto her, Thy brother shall rise again.

Martha saith unto him, I know that he shall rise again in the resurrection at the last day.

Jesus said unto her, I am the resurrection, and the life; he that believeth in me, though he were dead, yet shall he live.

And whosoever liveth and believeth in me shall never die. Believest thou this?

(John 11:23-26)

The resurrection of Lazarus is a narrative paralleling the resurrection of the "Redeemed" in the Rapture in the very near future. This is the prophetic side of the story that is often overlooked. Only by applying the prophetic *"after the second day"* concept, can you get the full meaning of the narration. With the application of the Six-Day Theory, in all of its forms, we are able to receive a deeper and richer understanding of the Scriptures.

THE MARRIAGE AT CANA

The wedding at Cana is the site of the first miracle that Jesus performed, by turning water into wine. This might seem to be a simple story of a man, a few of his followers, and his mother, who were all invited to a local celebration with friends and neighbors. The names of the bride and groom were not even mentioned, nor were any dignitaries noted in the account. With the exception of Jesus performing a miracle, the rural setting was quite charming, but ordinary.

> *"And there were set there six waterpots of stone, after the manner of the purifying of the Jews, containing two or three firkins apiece."*
>
> *(John 2:6)*

When we utilize the Six-Day Theory, we are able to delve deeply into this story for nuggets of gold. In John 2:6, the *"six waterpots of stone"* were filled with water, then Jesus turned the water into fine wine. Let's interpret the prophetic meaning of this portion of the story. The "six waterpots of stone filled with water" are the six thousand years of man being born into the natural world, living with hearts of stone. Biblically, "the water birth" is how we are born into this world, in water. For one to be born into the Kingdom of God, one must have his heart converted, from the water (natural birth), into a supernatural birth by the blood of Jesus (Spiritual birth) symbolized by the wine. The Bible calls this the second birth or being "born again."

The very first verse of this narration of the wedding feast gives us an immediate understanding that we may use the portion of the Six-Day Theory, which is called "On the Third Day."

> *"And the third day there was a marriage in Cana, of Galilee; and the mother of Jesus was there.*
>
> *And both Jesus was called, and his disciples, to the marriage."*
>
> *(John 2:1&2)*

The narration begins with *"And the third day there was a marriage"* which immediately tells us that we may apply the "On the Third Day" portion of the Six-Day Theory. The interpretation refers to a greatly anticipated event, called the "Marriage Supper of the Lamb." Again, this will take place after the "second day," or at the beginning of the "third day," just as was previously illustrated. This event will occur after 6000 years from Adam, or 2000 years after the birth of Jesus. After the Rapture occurs, about (AD2000), the dreaded seven-year Tribulation period of God's wrath will be poured out upon the unbelieving world. The "Marriage Supper of the Lamb" will take place at the end of the "Tribulation" period, which is "On the Third Day," or in the third millennium after the birth of Jesus.

> *"Let us be glad and rejoice, and give honor to him; for the marriage of the Lamb is come, and his wife hath made herself ready.*
>
> *And to her was granted that she should be arrayed in fine linen, clean and white; for the fine linen is the righteousnesses of the saints.*
>
> *And he saith unto me, Write, Blessed are they who are called unto the marriage supper of the Lamb. And he saith unto me, These are the true sayings of God."*
>
> *(Revelation 19:7-9)*

The marriage supper of the Lamb is one of the great promises of God that all Christians are looking forward to with great anticipation. All of the Redeemed, from the time of the creation of Adam, to the last person that will be saved out of the Tribulation period, will be there. The entire relationship of believers of the New Testament, to the Lord Jesus Christ, is based upon the ancient Jewish wedding rituals and customs. This is that portion of the ceremony in which the Bride and the Bridegroom are celebrating their wedding at a feast in his Father's house. This feast lasts for seven days (or years). All of the wedding party; the Bride, the guests, and the friends, have been invited personally by the Father. I wonder how many will be there at that glorious event? Have you received your invitation? Will you be there?

THE RESURRECTION OF JESUS CHRIST

The resurrection of Jesus is saved for the final example, because it is deeply meaningful and dear to the hearts of so many people. This powerful event has a myriad of meanings for the Christian world. The love and the power of God raised Jesus from the dead. The combined powers of Satan, the Roman Empire, and evil men, could not keep Jesus in the tomb. Jesus suffered the agonies of torture, the cruelty of crucifixion, and excruciating pain in paying for the sins of the world. He was hurriedly buried in a borrowed tomb; a large stone covered the entrance; an official seal of Caesar sealed the stone; and Roman soldiers were ordered to guard the tomb. All that could be done to keep Jesus in the tomb was done. For "two days," the enemies of Jesus thought that victory was theirs.

"On the third day," the loving Father came for His son. With an earth shattering roar, the earth quaked and shook; the soldiers fell as though they were dead; and the stone was rolled away. The power of Almighty God filled the tomb; and the Shekinah Glory of God radiated like a thousand suns. When God spoke, it was like the blare of a thousand trumpets and Golgotha shook: just like Mount Sinai rumbled and shook when God spoke with Moses. As God spoke to the dead, earthly body of Jesus, the earth trembled; and the body was immediately changed, in the twinkling of an eye, into a glorified living, heavenly, eternal body. The power and glory of God radiated from the glorified body of Jesus, which will never again know death. Jesus overcame death and the grave so that all who trust in him would have eternal life. The Sunday after the Passover Sabbath is the Jewish feast day, called the "Feast of the First Fruit." Jesus was the "First Fruit" of the Resurrection power of God. The "blessed assurance" of every believer is: that if God raised Jesus from the dead, Jesus has promised to raise everyone from the dead who trusts in Him.

The prophetic parallel here is: that just as Jesus laid dead in the tomb for "two days" because of sin, so has mankind been dead in our sins for "2000 years" since the birth of Jesus. Jesus paid the penalty for all the sin ever committed since the creation of Adam. His "six hours" on the cross represent the 6000 years from Adam to the Rapture of the Redeemed. But on the "Third Day," Jesus will come for His Bride, just as the Father came for His son. The blue skies will be pierced with a roaring shout, and a trumpet blare will awaken the dead in Christ; and the believers who are alive shall be taken up to be with the Lord forevermore. The earthly bodies, both of the dead and the living, shall be changed, in the twinkling of an eye, into glorified eternal bodies just like that of Jesus. This is the promise of the Lord God Almighty to everyone who will trust and believe on the Lord Jesus Christ. What better illustration of His promise could God give to us than the Resurrection of His own son.

> *"For if we believe that Jesus died and rose again, even so them also who sleep in Jesus will God bring with him.*
>
> *For this we say unto you by the word of the Lord, that we who are alive and remain unto the coming of the Lord shall not precede them who are asleep.*
>
> *For the Lord himself shall descend from heaven with a shzut, with the voice of the archangel, and with the trump of God; and the dead in Christ shall rise first;*
>
> *Then we who are alive and remain shall be caught up together with them in the clouds, to meet the Lord in the air; and so shall we ever be with the Lord."*
>
> *(1 Thessalonians 4:14-17)*

The scenarios that have been illustrated for the Six-Day Theory, are only a small sampling of the riches that may be gleaned from the Scriptures. The terms "After Six Days;" "After Two Days;" and "On the Third Day" are found throughout the Scriptures for you to enjoy in your own research. This is, perhaps, the most interesting of the theories that are directly linked to the Creation. It has been enjoyable sharing these rich illuminations with you.

X SUMMARY

The theories of the Creation, as we have discovered together, are numerous indeed. Many of these theories are quite humorous and even beyond reason; so as you may have imagined, some theories are not worthy to be mentioned. The one basic fact about the Creation of the Universe, or of Man, is that no man was there to witness and record what happened. If you believe that everything that exists created itself, that is your choice. If you believe that the Creator God created all of Creation, then that too is your choice. We have shared together many fascinating theories of man, but there is only ONE who claims to have created all things. He was not only there as the Creator, but as a witness to all of the creative events; and has left us a concise record of these events in the Holy Bible.

Let's continue our quest for knowledge and truth about the "Beginning of All Things." We have discussed man's theories about the beginning of all things. We have also discussed man's theories on Creation. Now let's seek the truth of God's account of his Creation. Since God said that He is truth, and that ALL knowledge begins with Him, let's pursue this truth through His Word. If we assume that God is the Creator, and the Biblical account is true, then we should have a very interesting journey experiencing God's narration together. There can be nothing more exciting then to explore and search for the beginning of ourselves and of our beautiful, vast universe.

Every journey has a beginning, but our journey will begin before there was a beginning. We shall journey back before the space/time continuum was ever created, a time when nothing existed. There was no universe, no solar matter, and not even time itself existed. We shall begin in eternity past when all that existed was God. Together, we shall take a most exciting *"Journey Back To Eternity."*

Chapter 3

And God Created

Have you ever wondered what it would be like to have some kind of supernatural power? We have all wondered, at one time or another in our lives, what would it be like to have X-ray vision; or perhaps, it would be exciting to have superhuman strength. There are so many of these powers that are reflected in children's comic books, or in the exploits of superheroes on television and in the movie theaters. There seems to be no end to the powers that are available to these superheroes. They are able to fly, become invisible, freeze or melt things with their laser-powered eyes, and much more. All of these concepts are exciting to us because we all have experienced these same ideas in our own imaginations. These cartoon-like characters are only the realization of our own fantasies. Our destination is to explore the *"beginning of all things,"* in the realm of everlasting eternity. It is within the realm of the eternal supernatural that we are about to travel together with our imaginations. This is the greatest adventure that man could ever embark upon, to go beyond the space/time continuum, and look upon the glory and the face of the God of Creation.

Literature throughout the ages is replete with great adventures into the unknown. Many of the adventures of fantastic travels are literary classics. Some of my favorites are "The Odyssey," "1001 Arabian Nights," "Gulliver's Travels," and "Pilgrim's Progress," just to name a few. My all-time favorite is "The Time Machine" by H. G. Wells. Wells created a machine in his imagination that could travel through time, either in the past, or into the future. His machine is the product of the nineteenth-century imagination, but the machine that we are about to travel in is of the twenty-first century. Our travels will take us not only through space and time, but beyond time itself.

As we enter our imaginary travel machine together, not only do we have the capability to explore time as we know it, but we are able to explore beyond the space/time continuum into the Spiritual Realm of eternity. Let's call our machine the "Eternity Explorer." In our machine, we will truly be able to go where no one has gone before.

Let's begin with traveling slowly back through the time of our own history, so that we may become familiar with our ship. Our first stop is earth's moon, as we anxiously await the Apollo lunar landing vehicle. Our excitement seems to be uncontrollable as the "Eagle" has landed. This is the greatest accomplishment that man has ever undertaken. It was the first time that man has put his footprint upon any planetary body other than Earth. Let's not tell them that we were there first.

Next, we have landed in the lush fields of Bethlehem. It is a cool fall evening, and the sheep are unusually restless. The patient shepherds gently calm their flocks, when suddenly angels appear overhead, and announce the birth of the Savior. The night skies become brilliant with the radiant glow of an angelic host singing praises to the newborn king. What a glorious sight! As our hearts throb with excitement and our eyes swell with tears, we begin to rejoice with the angelic host that fill the skies. Just as suddenly as they have appeared, the angels disappear from the skies, but the sounds of their joyous singing and praises still ring in our ears. This scene has always been a desire of my heart to witness, and in our Eternal Explorer craft it is quite possible to visit here any time we wish.

Now that we are learning to operate our imaginary travel craft, let's go back further in time to a desolate area of the Sinai Peninsula where it meets the Red Sea. Our destination is where a massive crowd of people is gathered at the shore. We shall join this enormous crowd of about two and a half million people, to observe and experience this great historic event with them. Fear has gripped the people as they have learned that the Pharaoh's army is closing in on them and their destruction is certain. But God has sovereignly intervened, by placing a pillar of fire between the Pharaoh's armies and the fearful Israelites to protect them. Moses climbs to the top of a small hill at the shoreline, attempting to calm the fears of the children of Israel. He shouts with a loud voice, and his words ring in their ears:

> *"And Moses said unto the people, Fear not, stand still, and see the salvation of the LORD which he will show you today; for the Egyptians whom ye have seen today, ye shall see them again no more forever."*
>
> *(Exodus 14:13)*

Moses then turns to face the Red Sea, and as he stretches forth his staff, in the power of the LORD, he commands the waters to part. With a mighty roar, the winds part the sea, mounting up huge walls of water on either side, leaving a very wide expanse of dry ground for the Israelites to cross to the other side. We, along with the Israelites, stand gazing in utter amazement at the awesome power of the Creator God. As the last of the Israelites are crossing the sea, God removes the pillar of fire, and the Egyptian charioteers continue their pursuit with anger and vengeance after the Israelites.

Moses again stretches forth his staff toward the sea and commands the waters to return. With a mighty thunderous roar the sea comes crashing down upon the six hundred chariots, and all the host of the horsemen of Pharaoh. The Israelites rejoice and praise Almighty God for their deliverance. We are there to witness the sovereign intervention by a loving God, by protecting the lives of His people. Israel is very special to Yahweh (Hebrew name of God), because Israel is the beloved wife of God (Isaiah 54:5).

I VISIT TO THE STARS

We are about to leave earth on our fantastic voyage, back in time and into eternity. As we leave the planet earth behind we begin to contemplate the marvelous technological accomplishments of man. We are overwhelmed as we recount our experience of the glorious angelic announcement in Bethlehem. To experience the awesome power of God first hand, as He parted the Red Sea, was an experience that we have only read about, until now.

As the Eternity Explorer passes through our solar system, let's take a close look at Jupiter: it has always fascinated me. As we approach Jupiter, we can immediately see that it is enormous in size, and has a vast array of colors, ranging from white to yellows, oranges, and reds. Jupiter is a kaleidoscope of swirling clouds moving at tremendous speeds three times faster than Earth's rotation, with a storm system that seems to rotate in the opposite direction of the main flow of clouds. The Great Red Spot, as it is called, is so big that three earth-size planets could easily span the distance across its opening. Three hundred earth-size planets would be required to equal the size and mass of this beautiful planet. Even this giant is really quite small compared to the stars of the universe, which we are about to explore. We are beginning to realize just how insignificant we are on our tiny planet when we view things from a galactic perspective.

As we leave our familiar planetary system, we are now able to see our galactic system, which is called the "Milky Way." What a breathtaking sight this is. It looks like a gigantic pinwheel rotating in space, and it contains billions of stars. We are able to see some stars which astronomers call "Giant Red Stars." The "Giant Red Stars" are the biggest stars known to science. They are so enormous, that if we superimposed the centers of our sun and a "Giant Red Star," the radius of the "Giant Red Star" would engulf the Earth in its orbit, which is 93,000,000 miles from our sun. It is certainly difficult to absorb or even comprehend such enormous bodies that fill the universe. My heart tells me: the greater the creation, the greater the Creator. The wonders and cares that we shared on Earth seem to pale and melt away as we experience the glorious handiwork of our Creator. It is said, "the earth was created for man; but the heavens were created for the glory of God." I could add nothing more than to say that we truly serve an awesome God.

> *"The heavens declare the glory of God, and the firmament showeth his handiwork."*
>
> *(Psalms 19:1)*

Let's now travel to other galaxies in our Eternity Explorer, and experience many new and exciting sights. Our imaginations are limitless, as well as our curiosity.

II THE SPACE / TIME CONTINUUM

Travel from one galaxy to another is quite effortless for us now, and we are astonished to see such brilliant displays of color. Some of the galaxies radiate a cool, light blue color while others shimmer in an eerie, lime green color. The magnificent diversity of God's creation totally engulfs our imaginations. We have experienced pulsating Quasars, shrinking Dwarf Stars, novae or exploding stars, and the Giant Red Stars. The Cosmos is ablaze with color and wonder. We have decided to observe some of the nebulae, or star clusters; and to study one in particular, called "The Horse's Head." This nebula is a star cluster that glows with blazing, rich reds and oranges, with a blinding white central core. It has a distinctive black void shaped in the profile of a gigantic horse's neck and head in the midst of this blaze of color. There are no stars or color in this huge black void, which gives it a unique eerieness. We have been absorbing the magnificent beauty of this nebula in the

Orion constellation for some time and then, suddenly, something very strange begins to occur.

The stars of the nebula all begin to shift in unison in one direction. Our ship, the Eternal Explorer, begins to move in the same direction as the entire nebula. We move very slowly at first, but we do notice an acceleration in our speed. After quickly checking our instruments, we find nothing that would account for this unusual occurrence. We prudently back away from the nebula, as its movement increases. The entire nebula is accelerating away from our location. We can now see other star systems moving rapidly in the same direction. Stars from behind us are streaking past us with blinding speeds, traveling enormous distances in a very short time. Hour after hour, we watch millions of stars soar past us, and our eyes never seem to tire of this spectacular sight. We are concerned that we may be in the path of one of these racing stars. As we search the space where these runaway stars might come from, to our astonishment, the space behind us was totally black and void. There is not even one star to be seen in that direction.

The scene before us is strangely reminiscent of a fireworks display with the exploding starburst patterns that burst into a huge sphere of dazzling colors. The only difference is that we are seeing that picture in reverse, with all of the sparkles returning to the source of its beginnings. We now understand what we are experiencing as we race after the last of the billions of stars that seem to be disappearing at a single point in space. All the heavens seem to be funneling into a spatial vortex: a place of origin. As we look around us, an eerie feeling grips us, because the vast expanse of space that was once filled with countless galaxies and nebulae arrayed in brilliant colors, is now void and empty. Total blackness engulfs us: there is no sense of space, depth, or distance; and we are alone.

The only source of light is the focal point where the stars are rapidly vanishing. The Eternity Explorer is racing toward that light, for we have reasoned together that what we are experiencing is the "Creation of the Universe," only in reverse. We had forgotten that our controls were set to go back in time ever since we began our journey. We have been experiencing the creation, or the birthing of the "Space/Time Continuum." The space/time continuum is where all matter exists. All matter moves in space through time. Matter can be measured with three dimensions, and it exists in the fourth dimension. The entire universe consists of matter moving through time.

We are traveling at full speed through the void of empty space as the utter blackness surrounds us, and we are engulfed in a void of nothingness. Our only reference to anything at all is the light source of the vortex, in which all the billions of stars seemed to disappear. Our anxiety level reaches a new high, when suddenly the Eternity Explorer is severely jolted and immediately the utter darkness is no more. We are bathed in a brilliant array of colors that appear to be fluid and translucent. Not only is there dazzling color, but glorious sounds of huge choirs of singing and praising voices. The marvelous sounds of praise are similar to the angelic host we heard in the night skies over Bethlehem at the birth of Jesus, the Savior. Apparently the universe, as we know it, no longer exists. The space/time continuum is no more, and we are now in a whole new dimension of existence. We have passed from the physical universe into the "Spiritual Realm."

III FROM TIME TO ETERNITY

We no longer have any anxiety or fears, because everything seems so serene and peaceful in this dimension. A sense of warmth comes over us; it is like coming home where only love exists. We obviously have left the dimension of time and space, and entered into Eternity. The "Physical Realm" has disappeared and we now exist in the "Spiritual Realm."

The "Physical Realm," which is also called the "Space/Time Continuum," does not exist in eternity. In our Eternity Explorer, we immediately notice that distance, speed, time, or even space, have neither meaning nor relevance in eternity. Although we are certain that the "Spiritual Realm" does exist in the "Physical Realm." With the amazing sights we have encountered on our journey still fresh in our memories, we can with assurance, proclaim that Almighty God himself and the angelic host are totally involved in the affairs of mankind. We have experienced the wonders of the nebulae, stars, and planets of the universe. We were thrilled at the joyous sight of the angelic host in Bethlehem. Also we were awestruck at the compassion and power of God: parting the Red Sea. All things within the universe are the LORD'S creation, and are under His control. It can truly be said that, the LORD reigns over everything that exists.

"...Who is the image of the invisible God, the first-born of all creation;

For by him were all things created, that are in heaven, and that are in earth, visible and invisible, whether they be thrones, or dominions, or principalities, or powers—all things were created by him, and for him;

And he is before all things, and by him all things consist."

(Colossians 1:15-17)

From our discussion and our own experience, we may conclude that all things were created by God, and that all things consist or are held together by Him. The angels themselves can exist in the Spirit Realm and in the Physical Realm, but they too were created. The universe and all that is within it is only a fleeting moment in realm of Eternity. The "Space/Time Continuum" is not part of Eternity, but only temporarily exists to accomplish God's plan. God does not live in Eternity. God <u>is</u> Eternity.

IV IN THE PRESENCE OF GOD

Our journey has brought us into Eternity. It is a place of unbelievable beauty and peace. The calming serenity penetrates to our very souls. The joy that we share is precious indeed, for neither of us has ever experienced such feelings or emotions in the Physical Realm. It is extremely difficult to explain these heightened emotions in human terms, because this surpasses the human experience. The sounds of the angelic host that surround us course through our very beings. The sounds of rejoicing, and the praising of Almighty God, resound from afar, and then another host of angelic worshippers near to us echoes the praises. We only had a very small sampling of this beautiful presence, in the night skies over Bethlehem.

These wonderful sounds and gorgeous fluid colors have become euphoric and all consuming to us. We have become completely immersed in what we are observing and feeling. We are totally enraptured by everything within Eternity, as if our very nature has been completely changed. Suddenly, without warning, the glorious sounds of angelic praising begin to slowly diminish, and then there is silence. The beautiful translucent colors cease their swirling motion, and then begin to rapidly fade, until there is only utter darkness that engulfs us once again.

There is only one source of light and we are being drawn closer to it with every passing moment. This is a radiance that I have never seen before. It is the purest white in color, and brilliant to look upon, especially with the surrounding blackness as a contrast. This unknown entity is of a nondescript shape, at least from our observation. Because of the glare of the radiation, it is impossible to discern any defining configuration or shape. As we are drawn closer to the light, we are again immersed in pure love. We have stopped at a respectful distance from the light, and we have become totally bathed in the golden, white radiance of the light, the Shekinah Glory of God.

Joy and peace now surge through our beings, even unto our very souls. We are swallowed up in the pure love of a Holy God. We are in the presence of Almighty God. Pure Love and pure Light emanate from God, as we rejoice in it. The Holy Scriptures say that *"God is Light,"* and that *"God is Love"* (1 John 1:5) and (1 John 4:8). The ultimate existence would be to live forever in the presence of God. We have no concept of the amount of time that we have been in the presence of God, because time has no meaning here. We have truly taken a *"Journey Back to Eternity,"* to the beginning of all things.

As we bask in the presence of God, we are able to hear three distinct voices emanating from the Light. Wonderful plans are being made: something about a plan for Creation and Redemption; and we thought that we heard our names being mentioned. The Scriptures state that God knew us before the worlds were created. It is difficult to understand this internal conversation, or perhaps it is not meant for us to have the complete full understanding of God's plan at this time. It is enough to know that we are part of the creative and redemptive plan of God: in this we may truly rejoice.

From the very beginning of our journey, our Eternity Explorer has been set to go backward in time into eternity. We cannot go any further back than to be in eternity, alone with God, before anything was created. We are truly *"In the Beginning with God."* Reluctantly, we switch the controls to go forward in time. Neither of us wants to leave the presence of God, but we must find out the answers to our quest. Since time has no meaning in eternity, we will just enjoy being here and let events unfold in God's own timing. Let's continue to be lavished in the continuous outpouring of God's love, and rejoice in the knowledge that we may experience God's love at any time. There is no end to this outpouring: it will continue forever and ever, from this eternity past to the eternity in the future.

V THE CREATION OF HEAVEN

As a flood of love pours over us, we become lost in His arms, basking in a state of love and compassion never before experienced. Our excitement builds, as we anticipate witnessing the beginning of God's creative plan to unfold. The Shekinah Glory of God suddenly begins to intensify, and with a sudden burst of sound like the roaring shout of a great waterfall, the outer darkness once again comes alive with brilliant, fluid color. Excitement fills our hearts, as this sudden creative act seems to awaken us from the euphoric rapture of being alone with God wrapped in His love. I don't know if God has ever been lonely in eternity, but it is logical to believe that if love is to be enjoyed, it must be shared. Since God is pure love, the entire creative and redemptive plan of God must be considered to be a wonderful love story.

THE THRONE OF GOD

We are immediately drawn to a site of extraordinary beauty. It is an enormous array of radiance, nearly equal to that of the Shekinah Glory of God. The Throne of Almighty God has been created by God Himself: it is truly a throne befitting the King of the Universe. Again we are in awe at the beauty of God's first creative act. No human words are eloquent enough to describe what we are seeing. The Apostle Paul mentions that he was caught up to the third heaven, but he said that it was unlawful for him to even speak of what he saw and heard (2 Corinthians 12:2-4). The

Apostle John, on the other hand, was taken up into heaven, and commanded to describe the very Throne of God and all of the surrounding wonders of heaven.

> *"After this I looked and, behold, a door was opened in heaven; and the first voice that I heard was, as it were, of a trumpet talking with me; which said, Come up here, and I will show thee things which must be hereafter.*
>
> *And immediately I was in the Spirit and, behold, a throne was set in heaven, and one sat on the throne.*
>
> *And he that sat was to look upon like a jasper and a sardius stone; and there was a rainbow round about the throne, in sight like an emerald.*
>
> *And round about the throne were four and twenty thrones, and upon the thrones I saw four and twenty elders sitting, clothed in white raiment; and they had on their heads crowns of gold.*
>
> *And out of the throne proceeded lightnings and thunderclaps, and voices; and there were seven lamps of fire burning before the throne, which are the seven spirits of God."*
>
> *(Revelation 4:1-5)*

This vivid description of the "Throne of God" is replete with prophetic symbols that need to be explained to appreciate fully the awesome regal sight that the Apostle John had seen.

- *"After this"* means after the Church Age is complete, as described in the previous three chapters of Revelation.

- *John represents* the redeemed New Testament Church.
- *"Come up here"* speaks of the Rapture of the Redeemed being taken up from earth to heaven (1 Thessalonians 4:13-18).
- *"Jasper,"* the stone of the tribe of Benjamin, the last tribe, who remained true to the Temple and to Judah when the tribes divided.

The Jasper is pure white, symbolic of the Righteousness of God.

- *"Sardius,"* the stone of the tribe of Reubin, the first tribe, the first born is heir to the inheritance. The deep red Sardius stone is symbolic of the redeeming Blood of Jesus Christ.
- *"The Rainbow"* is the everlasting symbol of God's judgment and promise of redemption of Man and of the Earth (Genesis 9:13-16).
- *"Emerald"* the stone of the tribe of Judah, lineage to King David and to Jesus, Son of God. The Emerald's green color means "life."
- *"The Twenty Four Elders"* are the Redeemed from the Rapture, Twelve tribes of Israel (O.T.) and the Twelve Apostles (N.T.).
- *"Clothed in White Raiment"* means they are pure and Holy in their glorified bodies: symbolic of righteousness (Revelation 19:8).
- *"Crowns of Gold"* means the Redeemed are Kings that share the Throne with Jesus, the King of Kings (Revelation 5:10).

As we glory in seeing the creation of the Throne of God, we are reminded of the eternal history it will have throughout the history of man and beyond time, into eternity. We have only touched on the prophetic implications of Johns' narration, because our focus is not on the prophetic symbols, but to experience the "beginning of all things" and examine its impact on our lives.

The Throne of God will be the focal point of the Kingdom of God. The Almighty Authority of the Universe will emanate from this point. It is staggering even to attempt to comprehend its full meaning. Let's follow the major historical events relating to the Throne of God, so that we may better understand what we have just seen created.

- The Throne of God is created before all things, and becomes the center of praise and worship when the angelic host is created.
- After the creation of the universe and earth, the throne of King David will be the earthly symbol of the Throne of God; and God proclaimed that the throne of David would live forever.
- Jesus died on the cross on Golgotha, was buried, and resurrected from the dead; he ascended into heaven and sat at the right hand of God. "At the right hand" means the Father gave God the Son all power and authority to rule from the Throne of God (Mark 16:19).
- After the Rapture, during the seven-year Tribulation Period, Jesus is crowned at his coronation as King of Kings and Lord of Lords, The Redeemed cast their crowns at his feet (Revelation 4:10&11).
- After the seven-year Tribulation, Jesus returns to earth with the Redeemed, which is called the Revelation. The Lion of Judah, Jesus, will rule the world from Jerusalem for one thousand years, as the King of Kings: and the Redeemed are the Kings that rule the world, and Jesus is the King over them (Revelation 20:4-6).

- After the Millennium, Jesus will judge the "Unrighteous of the Universe," man and angels. The Final Judgment of all sin will be from the Great White Throne of Judgment (Revelation 20:11-15).
- After all sin is removed from the Universe, the Earth is cleansed by fire. The Earth has been made Holy for the New Jerusalem to come down from heaven, and to be the eternal home for the Redeemed. Within the New Jerusalem is the Throne of God and of the Lamb, and the family of God shall abide there forever (Revelation 22:1-5).

What a marvelous history the Throne of God continues to have, from eternity past to eternity future. We have been reviewing its history while moments before, witnessing the thrilling creation of the "Throne of God and of the Lamb."

THE TEMPLE OF GOD

Our excitement soars to new heights as we observe the brilliance of the Shekinah Glory of God intensifying to a blinding radiance. Our ship, the Eternity Explorer suddenly jolts, as the voice of God once again speaks. The sound is like the blare of a thousand Shofar Horns sounding all at the same time. Suddenly, a glistening golden structure stands before us. Its beauty overwhelms us, and we can only gaze in awestruck wonderment. The intense, radiant glory emanating from the structure makes it difficult to discern distinctive features. It is a very solid structure, and yet, the interior radiance pierces through the structure as though it was sheer lace.

This is the *Holy Temple of God*, and the golden radiance that we see is the glory of God. We have concluded that no structure, no building, not even the heavenly Temple of God, is able to contain the glory of God.

Just as God created a Throne, from which to rule the "Kingdom of God," He now has established a center for a "Worship System." We are beginning to understand the plan of God, for we know that God the Father will give all power and authority to God the Son to be the "King of Kings" and the "Great High Priest." These two creations now establish the foundation for the implementation of that plan.

We are able to move in a little closer to notice two golden objects within the Temple. This scene is glorious to behold, for we are now looking upon the heavenly *"Ark of the Covenant."* Its golden radiance of pure Love and Mercy pierces our very souls. The lid of the ark is called the "Mercy Seat," traditionally the place where God the Father dispenses mercy and forgiveness. With the creation of the Temple, God's wonderful plan of Redemption has already begun. Our hearts surge with pure love we have for God, because we can plainly see: He first loved us.

As the Apostle John was taken up into heaven, God revealed to him all things that are yet to come. John recorded all the prophetic events that shall come to pass, and we are privileged to be able read of the future. One of the many exciting events will be that the Redeemed will be blessed to see the Temple of God and its contents, including the heavenly "Ark of the Covenant."

> *"And the temple of God was opened in heaven, and there was seen in his temple the ark of his covenant; and*

there were lightnings, and voices, and thunderclaps, and an earthquake, and great hail."

(Revelation 11:19)

The second item that we see within the Temple of God is a golden *"Altar of Incense,"* with embers of fire burning within it, and a sweet aroma of incense rising from the Altar. The swirling smoke of the incense symbolizes the "prayers of the saints" rising to the Father in heaven. The Altar of Incense will be used as the signal to begin the "Seven Trumpet Judgments," upon the unrighteous of earth, during the Tribulation period. John gives us a vivid picture of this account.

"And another angel came and stood at the altar, having a golden censer; and there was given unto him much incense, that he should offer it with the prayers of all saints upon the golden altar which was before the throne.

And the smoke of the incense, which came with the prayers of the saints, ascended up before God out of the angel's hand.

And the angel took the censer, and filled it with fire from the altar, and cast it upon the earth; and there were voices, and thunderclaps, and lightnings, and an earthquake.

(Revelation 8:3-5)

The "Altar of Incense" is symbolic of the Holy Spirit, who is the link between the believer and God the Father. The Holy Spirit, who indwells the believer, aids him or her to utter their innermost feelings in prayer and supplication to the Father. The prayers of the children of God ascend to the Father as a sweet smelling fragrance. Both of the items within the Temple of God are symbolic of God's plan of Redemption. As we search through the golden radiance of the Temple, we are unable to see anything else within the heavenly structure. It is a great mystery to us: why is there nothing else in the Temple?

Then great revelation was imparted unto us. Far into the future, God will establish the nation of Israel, and He will command Moses to build a Tabernacle in the wilderness that will be patterned after the heavenly Temple. The temple built by Solomon, the second post-exile temple, and the third Tribulation temple will all be patterned after the heavenly Temple as well. The "Wilderness Tabernacle" of Moses will have many more items in its makeup than does the Temple of God. The reasoning of God begins to unfold for us.

The surrounding *"White Linen Fence,"* of the tabernacle of Moses, will be as a barrier to separate sinful man from a Holy God. In the heavenly Temple, the glory of a Holy God permeates all things. In like manner, the *"Veil of the Tabernacle"* is not present in the heavenly Temple. The glory of God in heaven knows no boundaries, nor could any barrier stand. In heaven, access to God is free and open, to lavish in, and to rejoice in, forever and ever.

There are many items missing in the heavenly Temple that God will command Moses to build for the wilderness Tabernacle. The items of furniture that will be included in the wilderness Tabernacle are the "Golden Showbread Table," the "Golden Lampstand," and the "Roof Coverings" of red dyed rams' skins and badgers' skins, and the partition, called the "Veil." All of these items will be part of the decor of the wilderness Tabernacle and the earthly Temples.

God's reasoning is flawless perfection. The solution to the mystery is becoming clearer to us. There is a common thread running through all of these items. Let's explore them one by one:

- The *"Golden Showbread Table"* will have twelve loaves of bread on it, baked without leaven. "Golden" represents the Holiness of God. The "twelve loaves" represent the twelve tribes of Israel, and "without leaven" means without sin. The "Golden Showbread Table" is symbolic of Jesus, the sinless Son of God, who is the "Bread of Life:" The only way to Eternal Life (John 6:35).
- The *"Menorah"* or *"Lampstand"* will be a large stand of solid beaten gold, which will be the only source of light in the Holy place of the Tabernacle. The Menorah will have seven lamps and the center lamp* will be called the "Servant Lamp." Jesus, the "Servant of God," is the Light of the World," and will forever be the light that guides men into the Holy place of God (John 8:12).

* Historical Note: The Mishnah, the Jewish book of tradition, records that at the hour that Jesus died on the cross on Golgotha, "the servant lamp" suddenly went out in the Temple. The priests were unable to relight it for the next forty years, until the Roman General, Titus destroyed the Temple in AD70.

- The *"Roof Coverings"* of the Tabernacle will be made of rams' skins and of badgers' skins, that will be dyed red. This will symbolize the shed "Blood of Jesus" that will be the covering for ALL the sin of mankind, which is God's plan of Redemption (Hebrews 9:11-15).

- The *"Veil"* of the Tabernacle will be a separation between a Holy God and sinful man. When Jesus dies on the cross, the "Veil" of the Temple will be split in two by the hand of God. Access to God the Father will forevermore be open to whosoever will come, because of the finished work on the cross by Jesus, God the Son (John 3:16).

As we ponder this amazing scene of the Temple of God, as it was created, we realize that everything we know of the Tabernacle of Moses and the following Temples are far in the future from our point of observation. Nevertheless, we have the knowledge of our history and of the Scriptures to give us a greater insight to God's plan. I believe that God is fully aware that in our quest for understanding, He is revealing much to us, and giving us a deeper and richer insight.

The common theme, that courses through each of the items that will someday appear in the Tabernacle of Moses, is the picture of the redemptive work of Jesus. The heavenly Temple of God has none of the items listed above, because in the eternal Temple there is only the eternal promise of God for Redemption in the Ark of the Covenant, and the Altar of Incense for the prayers of the Redeemed. The Triune God has covenanted with Himself, and has established the plan of *"Redemption,"* through the sacrifice of Jesus, as though it has already been accomplished. God has proclaimed to Himself, knowing the beginning from the ending, to consider the future acts of *"Creation"* and *"Redemption,"* are as though they were already completed.

> *"In hope of eternal life, which God, who cannot lie, promised before the world began,"*
>
> *(Titus 1:2)*

We are now able to observe the exterior surroundings of the Throne and the Temple of God, which appear to be created upon a sea of glass like crystal. The sea of glass only amplifies the breathtaking, radiant beauty that engulfs our Eternity Explorer. As we now slowly drift around the Temple of God, something else becomes apparent to us. The items that will be outside of the Tabernacle of Moses are also missing in the heavenly Temple of God. The *"Laver"* and the *"Altar of Sacrifice"* are not here.

The exterior items of the Tabernacle of Moses will have a similar prophetic connection to the history of man as will the interior items. The *"Laver"* of the Tabernacle will symbolize a "cleansing of sin before the Lord." The Levitical priests will cleanse themselves before entering the Tabernacle. Everyone who will ever attempt to approach God must be cleansed. Under the "Dispensation of the Law," the Levitical priesthood will represent Israel, and must be cleansed before entering into the presence of a Holy God. Under the "Dispensation of Grace," the shed blood of Jesus will cleanse every believer from all sin. Every believer who will receive Jesus as his or her Lord and Savior will be cleansed and will be able to draw near to God at any time. Every child of God would be able to approach God openly and freely, without fear, even unto the very Throne of God.

The *"Altar of Sacrifice"* will be a place to sacrifice animals for the sins of Israel, under the Dispensation of the Law. Each head of household will bring an animal to the priest at the eastern entrance to the Tabernacle. There is only one entranceway to God. The man will place his hand upon the head of the animal, face toward the presence of God, and confess his sins. The sins of his household will be transferred to the animal, which will then be slaughtered as a sin offering. The animal's blood will be sprinkled at the foot of the Altar, and the sacrifice would be placed upon the fires of the Altar, for the atonement of sin. The sins of Israel will be covered for only one year. The "Altar of Sacrifice" will be only a picture of God's plan of Redemption. The fulfillment of God's plan will not be in the sacrifice of sheep and goats, but will be in the sacrifice of the Lamb of God.

During the Dispensation of Grace, the sacrifice will not be of animals, but the only sacrifice will be Jesus, the Son of God. The "Altar of Sacrifice" will not be a pit of fire, but will be an "Old Rugged Cross," on a hill called Golgotha. The blood of animals sprinkled at the foot of the Altar could only cover the sins of Israel for a season, but the blood of Jesus spilled at the foot of the cross, will blot out every stain of sin for eternity.

As we view, in awe, the heavenly Tabernacle, we have gained great insight into the mind of God as He pours out His wonderful revelation upon us. The glorious golden radiance of the Temple of God thrills our very souls as we attempt to drink in that which we can observe, and reflect upon the things of the future in the Tabernacle of Moses. We now understand why the Bible tells us of only two articles of furniture in the Temple of God. The other items in the earthly Tabernacle will be only symbols or pictures of what Jesus would fulfill in the future. The Triune

Godhead had already devised the plan of Redemption, and deemed it as though it was already completed, before anything was created, and we were there.

Almighty God has foreordained and predetermined a plan of *Creation* and *Redemption*, and the creation of the heavenly Throne and Temple is the foundation of that plan. The Throne of God will be the central focus of the "Kingdom of God," but as of yet, there are no subjects to rule, nor is there a Kingdom in which to rule. God has established the environment in which His elaborate plan will flourish.

In the same manner, the glorious Temple of God has now been established in heaven, but as yet, there are no worshippers. The "Worship System" has been established, but we do not understand all of the details at this time. We may presume that the Temple of God is the foundation of God's plan. No structure could stand without a proper foundation to build upon. This perhaps, may seem strange to our way of thinking, but God's plan is perfect. God's mind is always totally logical, down to the minutest detail, and every detail has meaning and purpose. We have seen that God sometimes gives insight and revelation into His plans and purposes, we just have to try to stay tuned in to the will of God, and draw close to Him.

Although, we have had many surprises throughout our journey, "Back to Eternity," there may be many things that we will be unable to comprehend, as we return to "Eternity Future." God has said, in His Word, *"For my thoughts are not your thoughts, neither are your ways my ways, saith the LORD" (Isaiah 55:8).* We may safely assume that our understanding is limited, compared to the mind of God.

As our Eternity Explorer, once again, slowly glides around the heavenly Throne and Temple of God, we continue to be overwhelmed and completely enraptured by the brilliant array of beauty and color. Heaven's aura of liquid radiance holds us spellbound, and we have become transfixed in the presence of God's glory. We realize that time has no meaning here, but we know our destination is to return to our own time in "Eternity Future." Our hearts leap for joy as we begin to anticipate that something new and wonderful is about to happen, and we will be witnesses to the unfolding of God's creative plan. Our excitement grows with each passing moment, as we ask each other, "what will God do next?" Could any thing possibly be any more spectacular than what we have already seen? What glorious wonder could excite us beyond our present state? Then suddenly, in all of our excitement, stillness settles over us. In great anticipation, we hold our breath; something is about to happen.

Chapter 4

The Heavens and the Earth

Silence grips our souls, as we seem to anticipate that something is about to happen. Our eyes never leave the magnificent glory of God that is before us. Even though we have experienced the awesome beauty of God's creation of heaven, the Throne, and the Temple, we can never take our eyes off of the exquisite radiance of the Shekinah Glory of God. The presence of God is like being cradled in His arms—we know that we are safe and secure in His love.

Suddenly, God's countenance begins to intensify. Is it love growing within God? We do not think that it is possible for God to exude any more love. Because God is love, whatever God is about to do will truly be wonderful. The warmth of God's love permeates our very souls until we are nearly consumed by its power. Our own mortality and frailty is apparent as our heart and pulse rates are at near collapse and our bodies and minds are totally and completely riveted upon the glory of God.

The silence in heaven is broken. In pure love, God speaks with softness and gentleness, and something miraculous happens. We let out shouts of joy and delight as we see groups of spirit-beings appear all over heaven. Everywhere we look we are able to see more and more groups, even far beyond our range of sight. We cannot contain ourselves. Rejoicing, we are overwhelmed at the scene before us. In the midst of all this awesome beauty and splendor, God has created millions of life forms: God has created spirit-beings to inhabit this wonderfully, glorious spiritual realm. God has just created the "angelic host of heaven."

I THE ANGELIC HOST

Gazing upon this vast multitude, we are awestruck by their sheer, magnificent beauty. Untold numbers of angels are scattered throughout the Spirit Realm, but we notice they are all being quickly drawn closer and closer to the Throne of God by some unknown force. That same force has been drawing us to God from the very beginning. The desire to be in the presence of God is in the heart of man as well as in angels. As the angels draw near, we are able to see there are many different kinds of angels. Their awesome presence thrills us as we see these newly created beings up close.

Words to describe this magnificent sight elude us once again. As we drift near to the Throne of God, we see many warrior-type angels that appear to be guarding God Himself. These are huge, mighty creatures, having great power and are wonderful to behold. Many of them appear to be messengers eager to do God's bidding, and to take His message anywhere at a moment's notice. As the great host of angels begin to crowd closer to the Throne of God. We see a glorious sight! The entire angelic host now beholds the face of their Holy God and Creator. The host is garbed in glistening pure white garments that reflect the pure righteousness of God. Their brilliant radiance overwhelms us. Drawn in from every direction, they slowly approach the source of heaven's greatest light and power: the Shekinah Glory of God. It exudes pure love, and it is this love that draws them.

The angels stop at a respectful distance from the Throne of God, and gaze upon Glory with utter amazement! As they approach the glorious radiance of God, they bow their heads with deepest reverence, and stand in silence before Majesty and Glory. The heavenly silence is deafening. We anticipate that something wonderful is about to happen. We do not move a muscle, for we, too, are in awe of the Majesty of God. Then, suddenly something begins to stir. We quickly search the vastness of heaven to see if we can locate the movement. We look toward the Temple of God and notice the multitude of angels that are in and around the Temple. These angels are arrayed in white raiment and have golden sashes about their waists. The aura of the angelic brilliance shimmers and dances. The Temple glistens like an enormous golden diamond or crystal with thousands of facets. Every Temple angel turns toward the Throne, and raises their hands in praise. Then the silence of heaven is broken as the Temple angels begin to sing praises, and to worship. Their sweet rich tones begin softly and gently. The host looks on, as the voices of the Temple angels rise and they fall to their knees with arms outstretched toward God, they sing:

"Holy, Holy, Holy, is the Lord God Almighty
Holy, Holy, Holy, is our Creator God who is above all

Honor, and Power, and Glory to the Holy One"

The Temple angels now fall prostrate, in worship, as the mighty angels about the Throne join in the chorus to honor their Creator. Tears well in our eyes as the wonderful sounds envelop our very being. Waves of spontaneous joy, move through the entire angelic host of heaven as they offer praises in song and worship to their Creator God. Their praise rises to a deafening crescendo, and we join in the celebration. All of the newly created life is rejoicing and celebrating that new life, and every created being is giving honor and praise to the One who gave them life. Should we do anything less?

The Spirit Realm now has an untold number of spirit-beings. Created life has begun. The only life ever to exist before the creation of the angels was God, the creator of life. The Creator is masculine in gender, and has created the angelic host in His image. Angels were created as servants, not as a counterpart. The angels are called by various names: "sons of God," "servants of God," "holy ones," "heavenly host," "heavenly beings," and "spiritual beings." These magnificent creations are the "servants of God," who were created to execute God's will. Angels are messengers, to implement God's plan of Creation and Redemption. The word "angel" comes from the Greek word *"anggelos"* which means, "messenger." Almighty God will use the created "spirit-beings" to fulfill His plan that is continually unfolding before us.

Angelic choruses of praise and worship, resound throughout heaven. Each angel now fully understands that he was created in love to serve God. Every angel knows that he will exist eternally. They love God because He first loved them. As far as the eye can see, angels are laying prostrate before their Creator. We have no idea how long their praise will last, because time is meaningless in the Spirit Realm. But suddenly the praise begins to abate, and as a body of one mind, the angels slowly come to their feet. A hush falls over the great host, and there is again utter silence in heaven. All eyes are focused upon the Throne of God. We draw closer as we anticipate something marvelous is about to happen.

Standing in awe before their Creator, each angel seems to know that they are about to witness something wonderful. God begins to speak gently and tenderly, and with great love. Suddenly, all of heaven gasps in amazement, for standing before the Creator are new beings: great and mighty "Archangels." Their radiant beauty is beyond description. Powerful in size, strength, beauty, and majestic glory, they far surpass the other angels. The heavenly host stands in awe before the archangels realizing that the Creator has established a ranking system within the Spirit Realm. Breaking the awesome hush in heaven, the Temple angels, once again, begin to praise God, and all the host of angels rejoice, as they sing praises and worship their God. The heavenlies resound with rejoicing and song. All of the angels, regardless of rank, give glory and honor to their Creator. They celebrate life and the love that created that life, and the privilege that it will be to serve a Righteous and Holy God.

As the angelic host comes to understand the purpose of their existence, their desire to be close to God grows stronger. The presence of God seems to empower each angel to please Him and to do His will. In their ecstatic exuberance, the entire angelic host is in harmony with the heart of God. Their praise and worship pleases God, and they are delighted to be able to please their Creator. Then, once again they are filled with anticipation. Again, a hush falls in heaven. Total silence swallows us up. Each angel is completely captured by wonder and amazement as they seem to know that Creator God is about to do something even more spectacular.

The countenance of God changes, as He tenderly speaks. The sweetness in His voice exudes pure love. In a spectacular flash of color, a magnificent figure suddenly appears before the Throne of God. A burst of color encircles the entire Throne area. Standing before God, is the most beautiful archangel of all of God's creation. Breathtaking to behold, he is the most powerful of all the angels and full of wisdom. His countenance almost overshadows the very throne of God. Wearing a mantle that appears to be made of many kinds of precious jewels, his glory radiates through it and the entire Throne area is engulfed in brilliant, dancing sparkles of color. This archangel is truly magnificent to look at, for there is no other like him in all of God's creation. His radiant glory is exceeded only by God Himself.

The angels even in the very back ranks see this final glorious creation of God and they stand frozen in place, daring not to move. Every angel seems to be in a state of awe, even the other archangels surrounding the Throne. Startling to look upon, this archangel and his radiant glory was so overpowering, you feel like you have to turn away from him, but it is difficult do so. He looks round about the onlooking crowd that is admiring his beauty, smiles with delight. He turns ever so slowly so that all can see his glory and beauty. It is obvious to the entire angelic host, that this archangel would certainly be the leader over all of them. He now looks upon the Throne and He that sits upon it. He looks upon the face of

God and smiles and reverently bows toward the Throne. God lovingly speaks in quiet, gentle tones to the radiant archangel that is standing before Him, and says; *"You shall be called "Lucifer" meaning "Day Star."* The archangel "Day Star" bows once more toward the Throne, and once again turns to the adulation of the angelic host. As "Day Star" raises his hands over his head, the angelic host shout praises to God to give Him glory for the great things He has done.

The heavenly host is jubilant and is celebrating the wonderful creative power of a Holy and Righteous God. They have been witnesses to the miraculous, and they exalt their Creator. Their exaltation reaches every corner of the Spirit Realm and in particular, the Throne area, called "Heaven." Singing and praises resound in the heavenlies, as Day Star still remains before the Throne with his arms upraised. Perhaps, he believes the praise and the adulation are for him. The Creator is observing all of the activity about him and is well pleased. God's glory appears to intensify, as we think we hear voices from the Throne, saying; *"Perfect, just as we had planned."* The exaltation becomes thunderous and does not appear that it will end anytime soon. Let's remain here by the Throne of God and observe the various kinds of angels in more detail.

II WHAT ARE ANGELS

People, throughout the ages, have been fascinated with angels. The very concept of God created super-beings is also frightening to many people, whose faith and understanding is extremely limited or nonexistent. Ignorance, superstition, art, and literature have shaped the world's view of what angels are. Much of the misconception of the Biblical reality of angels occurred during the Dark Ages when the reading of the Bible was forbidden, or even the possession of portions of the Bible meant torture and death. Therefore, artists and writers used their own imaginations to express Biblical truth. Mythological absurdities replaced the truth of the Word of God.

The concept, that angels are little naked babies, with wings, that fly around shooting little bows and arrows inducing people into love, is a very ancient myth and is popular even today. We have greeting cards with cute little cupids displayed everywhere as a symbol of love. You may think that it is simply harmless fun, but in truth, it is blasphemous and a Satanic lie from hell. If that sounds a bit harsh, let us examine the subject from its origin.

Ancient cuneiform tablets describe that Nimrod, the King of Babel, the great grandson of Noah, had died, leaving his kingdom to his wife, Semiramis. Queen Semiramis now claimed that she was the "Sun God." She then had an illegitimate son that she named Tammuz. She further claimed that Tammuz was a supernatural, virgin birth, and was the reincarnation of King Nimrod, the builder of the tower of Babel. Queen Semiramis commanded that her son Tammuz was to be worshipped as *"Savior,"* and she was to be worshipped as the *"Mother of God."*

The mythology is that Tammuz was hunting one day and was gored by a bull and died from the wound, but he rose from the dead three days later. The people of Babel made a holiday of his death, which, by our calendar falls on December 25th. They called the holiday, "Istar," which is where we got the name Easter. The people of Babel celebrated the resurrection of Tammuz by bringing colored eggs and erected trees on stands, which were eternal life symbols in that ancient society.

Babel is the origin of ALL perverted false religion, paganism, and idolatry. Satan knew of the plan of God from the Garden of Eden, and has tried desperately to thwart it ever since. Satan's plan to corrupt the bloodlines of man, with the fallen angels, failed when God destroyed the world with the flood. In Babel, Satan created the world's first state religion. Babylon, as it was later called, was the "Mother of ALL Idolatry," and it was the most ancient of the world's religious systems and spread throughout the whole world.

Throughout the ages, the perversion of the "Mother and Child" that was promised in the Garden of Eden, through the "Seed of the Woman," is prominent in all the major cultures in the history of man. Satan used the promises of God for the Messiah, and twisted them into pagan cults and myths. There are four specifics that Satan used to spread idolatry throughout the great cultures of the known world

1. "The Mother of God" *2. "The Queen of Heaven"*

3. "The Virgin Birth" *4. "Resurrection from the Dead"*

The ancient civilizations used all four of these specifics to establish their pagan, idol worship, and created an entire mythological system to accompany each goddess and resurrected god son. The Satanic deception is throughout the history of mankind. The list of these empires is as follows, along with their goddesses and gods:

• Babel	Semiramis and Tammuz
• Babylon	Semiramis and Tammuz
• Phoenicia	Ashtoreth and Tammuz
• Egypt	Isis and Horus
• Greece	Aphrodite and Eros
• Rome (Early)	Venus and Cupid
• Rome (Constantine)	Mary and Jesus

These empires all perverted God's promises of a Savior, and Rome, last of the great empires, accepted all of the pagan gods, from every part of it's empire. Even the Roman church adopted these four specifics under the Emperor Constantine. Notice the Greek and Roman goddesses, Aphrodite and Venus. Both were goddesses of love (lust). Their sons, Eros and Cupid, likewise were gods of love (lust), used as the focus of cultic orgies. These pagan mythological gods, somehow over the centuries, became thought of as angels of God. This is a blasphemous lie from Satan, who ever since his fall, has attempted to pervert and thwart God's plan. These imaginary mythological gods have somehow, in centuries past, been twisted into becoming the perverted creations of the Righteous Creator God of the heavenly host. Even today, the interest in angels is extremely high throughout the world.

We are bombarded with movies, television shows, pop songs, books, tabloid press, lapel pins, etc., that are about angels. The world seems to be willfully uninformed, as to what angels are truly like. I must give credit to many authors, who have diligently attempted to bring the truth about angels to a sadly misinformed world. We have grown accustomed to accepting the offerings of the media as truth, regardless of how bizarre it may seem. The old advertising adage is true: "people will buy into anything, if it is packaged right." Most of the known world, including the Judeo-Christians, believes that angels look like little "Danny Cupid" with his little bow and arrow. Even the symbol of a heart shape with an arrow through it universally means "Love." This is a sad commentary of our society. Have a happy "St. Valentine's Day," and don't forget that heart shaped box of candy.

There are many other misconceptions about angels that people have, that are very strange. One of the most popular beliefs is that when a child or an adult dies, they will turn into angels. People do not become angels. God created all of the angels that will ever be created, before the worlds were made. On the other hand, when people die, their soul and spirit will either go directly to heaven, if they are saved; or if they are unsaved, will go directly to hell to await the final judgment at the Great White Throne of Judgment. Those souls and spirits that have gone to heaven will return with the LORD, at the time of the Rapture, to receive their eternal glorified bodies. The angels of God will accompany the LORD to redeem the living and the dead. The "Redeemed" are the Bride of Christ, that has at this time been elevated to a position higher than the angels. People do not become angels, they become the "Children of God" to live forevermore with God, as a family that will be served by the angels. Remember that the angels were created to be servants of God, but people were created to be the redeemed "Family of God."

We have all heard of St. Peter letting people through the pearly gates of heaven after they die if they have been good enough in life. Let's correct these misconceptions right now. Nothing in that statement is true. All believers who accept Jesus Christ as Lord and Savior are saints, and go directly to heaven upon death. Heaven does not have gates of pearl. The New Jerusalem has twelve gates of pearl. The New Jerusalem is the home of God and His Redeemed Family that will descend from heaven. After the Great White Throne of Judgment, the earth will be cleansed by fire and made Holy, to receive the New Jerusalem. Peter has no say in who enters into heaven. In this Dispensation of Grace, all who are saved by the Blood of Jesus will enter heaven. Finally, it makes no difference how good or how bad one has been in life. You can never be good enough to be granted access to heaven upon death. Our ONLY passport is to repent of our sins and receive Jesus into our hearts and lives. Did you ever wonder where people get such ideas pertaining to their eternal lives?

Have you ever heard anyone say, "I sure don't want to go to heaven to be an angel sitting on a cloud and playing a harp for eternity?" It is a ridiculous statement to make; but nevertheless, rarely has man ever been rational about his eternal life. There is not one Scripture verse, in the Bible, to verify that angels play harps, or that they sit on clouds. There are forty-eight verses that mention harps, but not one ever states that an angel played one, on a cloud or any place else. The Bible does mention, some of the Redeemed in heaven played harps and other instruments, to give praises to the LORD, at His Coronation celebration (Revelation 5:8). This idea is probably a product of Middle Age artists. We could continue our discussion of what many people believe about angels, but let's search out what God says about the angels that He created.

THE CHERUBIM

To find the truth about angels, we must research the account of the One who created them. We must refer to the Word of God. We have established the fact that angels are not people. All angels are supernatural, celestial spirit-beings created by God. Angels were created to accomplish the will of their Creator. They perform a myriad of functions to support the Creative and Redemptive plan of God. Angels are majestic and exceeding in strength, and yet they do not dominate humans unless specifically ordered to do so by God. They have super-human intellect and wisdom, but are not omniscient. Angels were created holy, and like men were given the freedom of choice and free will. There are many characteristics of angels that we may discuss later, but for now let's focus on one rank of the angelic spirit-beings: the *"Cherubim."*

Cherubim are one type of celestial beings that are mentioned in the Scriptures. Cherub in Hebrew means *"one grasped or held fast."* The *"im"* is the plural ending in Hebrew, as in Cherubim. The Cherubim, as we have already discussed, are not little naked baby people with wings, but are mighty angels of God. The Cherubim are symbolic of God's Holy presence and unapproachability. They are the spirit-beings that guard the righteousness of God. The first mention of Cherubim in the Word of God is in (Genesis 3:24), when God placed Cherubim at the eastern side of the Garden of Eden, to prevent sinful man from reentering the Garden to eat from the "Tree of Life." If Adam, or his family, were able to eat from the

Tree of Life, in their sinful state, then sin could live forever, which would be contrary to God's plan.

The most memorable use of the image of Cherubim is on the lid of the "Ark of the Covenant," which is called the "Mercy Seat." There were two golden Cherubim, who were facing one another, bowed down, with their wing tips touching. Golden figures of Cherubim were also on the "Veil" of the Tabernacle of Moses. Both uses of the Cherubim were to protect the Holiness of God. The function of the Cherubim seems to be primarily that of guardians of God. First, in the Garden of Eden, then protective images in the Tabernacle, as well as the Temple of Solomon, the second Temple, and the end-time Temple described in (Ezekiel ch. 41-44).

The actual appearance of Cherubim is extremely vague. According to 1 Kings 6:24, their description is that they have two wings, but in Ezekiel 10:21, the Cherubim have four wings, which could indicate individual ranking. The first chapter of Ezekiel is extremely colorful in his description of Cherubim. There is so much prophetic imagery used to describe the appearance of Cherubim, that it is impossible to piece together an accurate picture. The usage of prophetic terms such as: "like," or "likeness," or "the appearance of," is repeatedly used, along with vivid descriptive word imaging. Thus, it is left to our imaginations to decipher. Since the prophets of God could not find adequate words to describe the Cherubim, it seems doubtful that we could improve upon their attempt.

THE SERAPHIM

The origin of the name *"Seraphim"* is a bit vague. It comes from the Hebrew, and may mean "princes" or "nobles." There is very little Biblical information about the Seraphim. Chapter six of the book of Isaiah gives us the best picture of the function of the Seraphim. The Seraphim are definitely angelic beings, perhaps even similar to the Cherubim. From Isaiah 6:2, we are told that each Seraph has six wings, a face, hands, and feet. Two wings covered his face, two wings covered his feet, and he flew with the other two wings. Perhaps the use of the covering wings was to display humility while in the presence of God. We also know that the Seraphim speak to people, and when one spoke to Isaiah, the door posts of the house shook. Seraphim sing praises to the LORD, from above the Throne of God.

"And one cried to another, and said, Holy, Holy, Holy,
is the LORD of hosts; the whole earth is full of his glory,
(Isaiah 6:3)

The Scriptures also tell us that the Seraphim give service in the Temple of God, because at the command of God, the Seraph using a pair of tongs, took a live coal from the Altar, and placed it upon the lips of Isaiah, to symbolically purge his sin (Isaiah 6:5-7).

We can summarize this scant information with the authority of the Scriptures. Seraphim are heavenly spirit-beings. They are angels related to the Cherubim, and seem to be messengers of God. They can fly, and have access to the Spirit Realm and the Physical Realm. They have six wings, and have humanoid features, such as a face, hands and feet. The Seraphim are attending servants at the Throne of God, and at the Temple of God. They are messengers of the LORD, and relate His will to the prophets of God. Seraphim also worship their God, and sing praises unto the LORD.

THE ARCHANGELS

The third category of angels is called *"Archangels."* These angels are, perhaps, the most elusive of all the angels in the Scriptures. Archangel is a term that is only used twice in the Bible, but yet we know more about them than we do the Cherubim or Seraphim. The term "Archangel" is from the Greek word *"Archaggelos"* which means "chief messenger."

There are only three archangels mentioned by name in the Bible: "Gabriel," "Michael," and "Day Star," whom we call "Lucifer." Michael is the only angel that is identified as an Archangel, although Gabriel and Lucifer certainly qualify as the "chief messengers" of God. In fact, Lucifer could be defined as a super-archangel, far superior to all other angels. Let's examine these super spirit-beings a little closer, and see if we can understand what exactly are these creatures, and what is their function.

MICHAEL

Michael is the easiest to identify as an archangel, because the Word labels him as one. The name "Michael," in the Greek means *"who is like God,"* a worthy name for a messenger of God. He first appears to Daniel as a messenger from God, to encourage Daniel and to give him strength to endure. Michael also came to do battle with the angels of darkness that were trying to hinder God's plan. Daniel prophesied that in the "last days" Michael would stand up for the Jews.

The next time we see Michael, he is contending with the Devil. It appears that he is restrained as a warrior, perhaps under orders not to do battle, but to stand his ground against the Devil.

> *"Yet Michael, the archangel, when contending with the devil, he disputed about the body of Moses, dared not bring against him a railing accusation, but said, The Lord rebuke thee."*
>
> *(Jude vs. 9)*

Although he appeared to be restrained in doing battle with the Devil in the previous encounter, Michael, the warrior archangel, surely proves himself in the battle for the control of heaven. Michael's deeds are truly mighty and heroic, as befitting a warrior angel of God.

> *"And there was war in heaven; Michael and his angels fought against the dragon, and the dragon fought and his angels,*

And prevailed not, neither was their place found any more in heaven.

And the great dragon was cast out, that old serpent, called the Devil and Satan, who deceiveth the whole world; he was cast out into the earth, and his angels were cast out with him."

(Revelation 12:7-9)

GABRIEL

Gabriel is known as the "messenger archangel." Although he is never called an archangel, he performs all the duties of one. The name "Gabriel," in the Greek, has the meaning: *"God is Mighty."* Gabriel also was a divine messenger to Daniel, to explain the prophetic visions that Daniel had. Daniel was given an enormous amount of prophecy pertaining to the "last days," as well as vivid historical accounts of the future Greek Empire, as revealed by the archangels.

As the "messenger of God," Gabriel announced to Zacharias, who was a priest in the temple of God, that his wife, Elisabeth would give birth to a son, and his name shall be called John. This John was to preach repentance to the people, and to baptize believers. He became known as John the Baptist, who was to prepare the way for the Messiah. This was a very important message for an angel to give, and I believe that it was a task for an archangel. As important as that announcement was, it could not compare to the message that Gabriel was about to give to a young virgin, named Mary. The annunciation of the birth of the Son of God, the Messiah, is the centerpiece of the plan of God, for the Redemption of mankind. There could be no greater message given from God to man than this message.

> *"And the angel said unto her, Fear not, Mary; for thou hast found favor with God.*
>
> *And, behold, thou shalt conceive in thy womb, and bring forth a son, and shalt call his name JESUS.*
>
> *He shall be great, and shall be called the Son of the Highest; and the Lord God shall give unto him the throne of his father, David.*
>
> *And he shall reign over the house of Jacob forever; and of his kingdom there shall be no end."*
>
> *(Luke 1:30-33)*

This is a great honor for Gabriel, to deliver the message: the Son of God, the Savior of the world, was about to arrive on planet Earth. The news spread all through heaven, and the angelic host must have triumphantly rejoiced, as the Son of the Most High would now become incarnate. It was a wonderful announcement that Gabriel gave to the young girl, Mary, but the angelic host of heaven could not be contained at the birth of Jesus, for they filled the skies over Bethlehem that one fall night. Over three hundred and fifty prophecies have been given to Israel by the prophets of God about the first coming of the promised Messiah. The angelic host declared the fulfillment of every one those prophecies to the amazement of shepherds that were keeping watch over their flocks that fateful night.

"And suddenly there was with the angel a multitude of the heavenly host, praising God, and saying, Glory to God in the highest, and on earth peace, good will toward men."

(Luke 2:13&14)

LUCIFER

As fascinating as the stories are of the archangels, Michael, and Gabriel, there is none more fascinating than the greatest of all the angels, the archangel called, *Day Star*, who is better known to us as *Lucifer*. He was created to be the greatest of all of the angels, but because of his pride, and envy of God, fell in disgrace. Lucifer's goal, from the time that sin was found in him, was to thwart God's plan of Creation and Redemption. Lucifer's ultimate goal is to be god over the angels and over all mankind. We shall not expound upon his deeds at this time, but we will follow the actions of Lucifer as we observe the events of God's plan unfolding. We shall observe how sin entered the universe, and Lucifer's attempt to disrupt the plans of God.

The Scriptures are replete with his disruptive tactics. We see his deeds even today and into the future, but his time is short. We can read of his future exploits, but we have the assurance that he will loose. The Bible tells us about the beginning of all things and the end of all sin in the universe. The LORD is the Alpha and the Omega, beginning and the end, and He has revealed all things to us in His Word. Therefore, we can know the future that God has planned for us, in spite of what Lucifer, the archangel, may do to us, or how he may try to deceive us. His goal, at this time in the history of man, is to deceive us away from God, and then destroy and kill us. As this "Age of Grace" rapidly comes to a close, the Devil is furious, because he knows the time is near for the LORD to return for His Bride. We know that there is nothing that he can do to prevent the Rapture from occurring. We have called this archangel, *Day Star*, or *Lucifer*, but he has been called by many other names in the Bible, such as: "the Devil," "Satan," "Dragon," "Serpent," "Murderer," and "the Father of Lies," and many more names that he has earned. We certainly will pay close attention to this archangel, as we now return to observe the angelic host, which is still giving praise to God, and worshipping their Creator. Let's draw closer to the Throne of God, and become a part of this extraordinary angelic exultation.

III GOD REVEALS HIS PLAN TO THE ANGELS

Caught up in the Spirit of God, the host of angels has crowded close to the Throne of God in total ecstasy. The archangel Day Star has moved close to the Throne, finally realizes that the praise and worship was not for him, but for the Creator God. The Temple angels, arrayed in glistening white robes, and golden sashes, are lying prone toward God, offering their exaltation. The mighty angels that surround the Throne are on bended knee, with heads bowed, anxiously waiting to be obedient to the will of God. Choruses of angels are worshipping their Creator, and heaven resounds with glorious, beautiful harmony that seems to cascade from all parts of the Spirit Realm. Roaring like a gigantic waterfall, tumbling down from a high mountain, wild and carefree, unrestrained in its power, the angels' music majestically soars, then suddenly, is hushed, as all of heaven becomes silent. We actually feel the crush of the silence. Then, we too, noticed what the entire host is feeling as they begin to sense the will of God. Something wonderful is about to happen. The angelic host is enrapt in awe, as Creator God rises up from His Throne, and stands upon the glassy sea. Facing them, as His radiant glory fills all of heaven, God begins to address them for the first time.

> *"My beloved children, I am well pleased to see you so happy. We have created you in love, and you have already learned to respond in love. There can be no greater love, than to serve your God. You were created to serve, and in serving, you show your love to me. I am your Creator God, who loves you; there is no other like me. In ages past, I took council with myself, for there is none greater, to devise a plan for all eternity. It was predetermined that you would be an integral part of our plan. You were created in the image of your God, eternal spirit-beings that have a will and a choice to serve your Creator. Will you serve your God, and share in the greatest adventure that could ever be devised?"*

The angelic host is ecstatic, they rejoice, seemingly as in one accord. Among the myriad of angels, some perhaps, want to hear more before committing to such a venture, or maybe they are just confused, but nevertheless their consternation does not go unnoticed by the ever watchful Day Star, the archangel. The excitement in heaven is enormous. Temple angels were praising and worshipping, as others are excitedly discussing the words they heard, while the mighty angels around the Throne are anxious to hear more about God's plan. In the din of the angelic stir, God raises His right arm for attention, and an immediate hush falls over the host of heaven, as their God begins to speak once more.

> *"My beloved ones, the plan that I speak of is one of Creation and Redemption. I have given ALL power and authority, to "God the Son" (Matthew 28:18). My Son will do all the things that I have commanded, and My Words shall He speak, and My Will shall He do. The Son of God shall be the very embodiment of My Word. He shall be called the "WORD" (John 1:1). The Father is in the Son and the Son is in the Father (John 14:10). The Son and the Father are one (John 10:30). In all things obey the "WORD."*
>
> *God the Son shall create a completely new realm, and it shall be called the Physical Realm. The host of heaven shall be able to exist in the Physical Realm as well as it does in the Spirit Realm. It will be an environment of space; and physical matter will move in a span of existence, called time. It shall be wonderful to behold, a space filled with color and beauty in a whole different dimension. This array of fire and color shall reflect the Power and Glory of your God, and your Creator. Within this Physical Realm, there shall be countless fiery balls of*

gases and elements, that shall be awesome to observe, even for the heavenly angels. These fiery orbs shall be called stars, and they shall be without number. Not far from heaven, will be a system of planets that will orbit one particular star. One of those planets will be a very special creation, and shall be called Earth.

Planet Earth will be the home of the greatest of all our creations, which shall be called MAN. Man will be a race of mortal beings, that will be created in the 'Image of God,' just as the angels were created. These mortals shall have eternal existence, and will have an individual will and the power of choice, to either follow or reject their own Creator. Their mortal life will be but a brief span of existence, that will be filled with obstacles, trials, and temptations to overcome, throughout their existence.

The Son of the Most High will provide the way for all mankind to live forever, in the presence of their Creator (John 14:6). Those mortals, that receive My Son, shall be redeemed, and shall live by My Word, and in the power of My Spirit. My Spirit shall dwell within them, and shall seal them for that promise of eternal life, that will be given by the WORD (Ephesians 1:13). These mortals shall be frail, and have neither might nor power of their own to overcome, but they shall overcome by My Spirit (Zechariah 4:6). To every mortal that overcomes, to him will I grant to sit with Me and My Son on the Throne of their God, and your God (Revelation 3:21).

Every mortal that will trust in the 'WORD' shall be redeemed at an appointed time. I have established this appointed time to redeem 'My Children' out from the Physical Realm, to be caught up into the Spirit Realm of heaven, to live forevermore. The corruptible shall put on incorruption, and the mortal shall put on immortality, and they shall see their God (1 Corinthians 15:53). Their Redeemer will prepare a magnificent home for My beloved Children. The Redeemed shall be Our Children, Our Bride, and Our Wife. Out of all of the beings that I will have created, My beloved 'Redeemed' shall be 'My Counterpart.' The 'Counterpart' of your loving God will be the only feminine entity in existence. You shall serve My Counterpart, as you serve us, for she too shall be high and lifted up. We shall all be a family, who will be created in love, and shall be sustained in love, forever and forever."

The Creator then majestically returned to His Throne, and His glory appeared to radiate even more, if that were possible. God had just shared His plan of Creation and Redemption with the angelic host, and they are frozen in place, absolutely stunned at the revelation. The mighty angels around the Throne, suddenly, as if of one mind, gather in front of their Creator and bow; then, as a unit, kneel before their God, to await His command. Once again, the Temple angels begin to sing and praise their God and Creator.

"Holy, Holy, Holy is the Father
Holy, Holy, Holy is the Son
Holy, Holy, Holy is the Spirit"

Immediately, the entire angelic host, as if suddenly awakened, begins to praise and worship God. Some angels excitedly grasp one another, in jubilant conversation, while others are jumping up and down with excitement. Most of the angelic host falls prostrate toward their Creator in complete obeisance and worship. What a marvelous sight to behold, and yet, as we examine the host more closely, many of the angels appear to be either stunned, or do not like what they heard. The most obvious is the archangel, Day Star. He appears to be deep in thought, and sullen in his demeanor. What is this beautiful, mighty, archangel thinking about? Well, we are not going to concern ourselves with Day Star at this time. Let's rejoin this great celebration. This panoramic scene before us is beyond our senses to fully absorb. The sound is overwhelming, and the majestic glory of God is the light of all of heaven. The host of angels is without number, and their power would seem to be limitless. And yet, with all that is before us, we believe that greater wonders are yet to come. Even as we speak, God is beginning to rise from His Throne, perhaps in appreciation of the angelic praise, or perhaps the Creator is about to do something new and wonderful.

As God stands before His magnificent Throne, the deafening praises begin to subside, and then once again, an absolute silence encompasses all of heaven. The mighty angels guarding the Throne make a pathway straight out from the Throne, like a military honor guard, lining both sides. God then begins to walk along this aisleway, His radiant glory shining as never before. Suddenly the Creator God stops, and within that brilliant glory, a figure emerges and continues walking. The angelic host gives a collective gasp, startled and amazed. They see before them two figures both with equal glory, power and majesty. One is not diminished by the other, but they are in total harmony. "God the Father" returns to the Throne and as He sits down, He points toward the standing Glory, and says, *"This is My beloved Son, in whom I am well pleased."* The "Son of God" turns and faces "God the Father" and bows down and kneels in reverence to His Father.

Led by the Temple angels, the entire stunned angelic host begins to softly and reverently worship and praise "God the Father" as well as "God the Son." The "Son of God" rises to His feet, and in all His glory and splendor, turns from "God the Father," and slowly walks away from the Throne, in the middle of the angelic honor guard. With purpose and determination evident in every stride He takes, the Son of God quickly walks upon the glassy sea. As He arrives at the end of the angelic guard, He again turns toward His Father, and bows. The entire angelic host is tensed with excitement, knowing that something wonderful and marvelous is about to happen.

Chapter 5

The Earth Was Without Form and Void

The angelic host is in awe of the Son of God, as He walks to the end of His honor guard, in front of the Throne. His radiant glory is majestic, and wonderful to behold. His loving gentleness is like that of a "lamb," as He once again lovingly turns toward His Father, to give honor to Him. His countenance is set like chiseled granite upon His face, and His movements cadenced to perform this predestined event, without hesitation. The WORD turns away from the Throne, and raises His arms over His head and gives a mighty shout. Like a mighty, majestic, "regal lion," His command roars through the canyons of Heaven, and into every part of the Spirit Realm. Heaven shakes violently, and the angelic host is terrified. Instantly, all of Heaven is encased in darkness, as far as we can see. Utter darkness surrounds us in every direction, and yet Heaven itself is just as brilliant as ever. Still stunned and terrified, an eerie silence falls on the heavenly host. Even the mighty angels around the Throne seem to draw closer to their God, to seek comfort and protection, in this time of personal anguish, and fear of the unknown.

We also are recovering from the initial shock, and our ears are still ringing from that roaring shout. Possibly, we can offer some plausible explanation, as to what has just occurred. We witnessed everything, and yet we are only able to apply a simple analysis of a very complex event. Suppose you had a large balloon, and filled it only half full of air. Extend your finger of one hand, and begin to press your finger into the soft balloon, while holding it with the other hand. You will be able to move your fingers around anywhere inside of the balloon, without being actually inside of the balloon. Imagine the inside of the balloon to be the created darkness of "Space," or call it the "Physical Realm," if you like. Everywhere else, outside of the balloon, is the "Spirit Realm." The "Physical Realm" exists within the vastness of the limitless "Spirit Realm." As vast as the "Physical Realm" may seem to us, it is still only balloon sized, when compared to the eternal, never ending vastness of the "Spirit Realm."

As we are observing the utter blackness of "Space," we notice that the darkness did not have any effect on the "Spirit Realm." We, who were outside of the balloon, in the "Spirit Realm," are not effected by the laws that govern the "Physical Realm." We were also wondering, why didn't the darkness overtake us, since it appears that the darkness was all around us? Since our attention was riveted upon God the Son, standing there in all of His majesty, and glory, it became immediately obvious to us. *"God is Light, and in Him is no darkness at all"* (1 John 1:5). The darkness had to remain where the WORD had commanded it to be, because darkness cannot exist in the presence of the Light of Almighty God.

I THE CREATION OF PLANET EARTH

The WORD, pointing His arm with a deliberate calculated aim, pointed to a position near to heaven. With a gentle, loving, command, there appears below us a beautiful blue globe. This has to be planet Earth, with its soft colors of blue, tan, green, and white, gently swirled together. It appears to glisten in the light emanating from Heaven. Remaining in their stunned frozen position, the angelic host dares not to move. No one speaks a word, as again complete silence reigns in Heaven, every angel watching the WORD intently.

Unexpectedly, and without warning, the WORD begins to ascend above the glassy sea, and slowly descends toward the brilliant blue planet. The Son of God entered into the black empty space of the Physical Realm. It seems that Heaven acts as a portal between the "Spirit Realm" and the "Physical Realm" of Space. As the WORD enters the darkness of Space, His radiant glory illuminates the emptiness, and the darkness has to flee. We all watch as the Shekinah Glory of the Son of God travels all the way to the blue planet, and then disappear behind the planet. Within moments, the glory of God begins slowly enveloping the entire planet. The WORD has filled the whole Earth with His glory, and made it Holy. The whole Earth is filled with the radiant fire of God, and yet is not consumed, perhaps much like the fiery bush that Moses saw. Planet Earth glistens and sparkles like a gigantic jewel, set against the black velvet backdrop of Space. The WORD has placed the "signature of God" upon planet Earth, and His holiness is prominently present in the darkness of Space.

Let us take a moment here to analyze what we have just witnessed. The creation of the Physical Realm and planet Earth, is a spectacular event. We earlier discussed the "Gap Theory," that states: "the Universe was created long before the creation of the Earth." In all fairness, we are now observing the scenario where the Earth is being created first. This scenario is not without merit, and should be explored. As we await the return of the WORD, we have time to research this theme.

In Genesis 1:1, we observe that *"In the beginning God created the heaven and the earth."* The Hebrew word for *"heaven"* is *"Shamayim,"* which is always written in the plural, and always is in the masculine gender. If the word *"heavens"* (pl.) is used, one could assume that God is speaking about the original creation of the whole universe, including the earth. One could also assume that since the universe, (heavens) and the earth are mentioned separately, then it is reasonable to assume that they were created separately, thus scholars have established the Gap Theory.

If the word *"heaven"* (singular) is used, we may then assume that this is not universal creation, but is only the creation of earth and its atmosphere, which the Bible calls "the first heaven." To further support this scenario, the sun, moon, and stars of the universe, "the second heaven," are not mentioned until the fourth day of the recreation of the earth (Genesis 1:14-19). In these verses, the Bible is speaking about God allowing the lights of the moon and stars to shine in the darkness of night. If the word *"heaven"* is used in the singular, then, we can conclude that the original creation of the universe is never mentioned, but only the recreation of planet Earth. Therefore, we may also conclude that *"In the beginning"* is the original creation of only the earth in (Genesis 1:1) and the recreation of earth in (Genesis 1:2). It is the beginning of planet Earth and nothing else. Herein lies the problem for Bible scholars, because if *"Shamayim"* is always written in the plural, when can it be used in the singular? We are not able to resolve this linguistic mind bender at this time, so I have presented both sides of a very interesting academic problem.

Returning to our heavenly scenario, we can observe that the WORD is returning to heaven, as His glory now, again, lights up the darkness of Space. There is something very strange here. When we look down at the earth from our Eternity Explorer, it is a very long distance from the edge of heaven. When we cruise over the glassy sea of crystal, we are able to see, close up, every detail of earth, as though we were only a few hundred feet above the surface of the earth. The WORD has created heaven and earth with such precision and accuracy, that everything that occurs on earth is observed in heaven. This is marvelous. It is just another revelation that God loves us so much that He cares about everything that happens to us, His precious creations. We all can take great comfort in knowing that our Creator is a loving and caring God.

II THE CREATION OF THE UNIVERSE

All the angels of heaven are intently watching the WORD in His radiant glory soaring through the utter darkness of Space, lighting it, as He returns to heaven. The Temple angels once again begin to softly sing praises to the Son of God, while the remaining angelic host remains silent. Then the WORD soars over the awaiting host. He lands before the Throne of His Father, and as He points toward the earth, proclaims in a loud voice to the entire angelic host.

> *"Be thou exalted, O God, above the heavens; let thy glory be above all the earth."*
>
> *(Psalms 57:5)*

The Father arises from the Throne and speaks gently to the Son. *"My Son, you may proceed with the plans that we have predestined before all things were created."* While the Father and the Son are speaking together, the radiant glory of both intermingle and intertwine as though they are of one body and one mind. It is a very tender moment to behold, much like a warm homecoming for the Son. Since we are still in heaven, we do not know how long the WORD was away on planet Earth, because we are still in the Spirit Realm, where time has no meaning.

The WORD finishes speaking with His Father, turns and walks to the same place where He created planet Earth. We are still amazed at the sight of the glory of God separating into two glories, into two individual entities, and yet one would perceive that they were happier when they were together. The Son's shining countenance is one of determination to do the task set before Him, in obedience to the will of the Father.

Without hesitation, the WORD stretches forth His right arm toward the planet Earth. He commands a sun to appear in the distance and a brilliant light appearing very small at first and then growing to an enormous size, then seems to diminish in size and brightness as it travels at a tremendous speed away from us. The star comes to rest exactly where the WORD has commanded. It lights up the darkness of Space, and illuminates the earth, like a dazzling precious jewel. A moon orbits the earth reflecting the warm radiance of the star. We are now able to observe many other planets that were also commanded to orbit this beautiful star, which we call the sun. We have visited some of these planets before, but now we have witnessed their creation. This truly is an amazing glorious sight to see. The entire planetary system is operating exactly as the Creator had designed it. Each planet has its own size, color scheme, rate of rotation, and chemical makeup for individuality. Although every planet has its own beauty, none is more beautiful than planet Earth. Earth is the only planet that has the glory of God upon it.

The WORD waves His arm across His body like a sower casting seed upon the ground, and gives a thunderous shout like the roar of a kingly lion. All of heaven shakes at His mighty shout, and the roar echoes into every corridor of heaven. Terrified angels draw close to one another. A ball of fire appears, quickly growing to an enormous size with a blinding brilliance emanating from it. With a mighty "big bang," this fiery mass then explodes into millions of fireballs, which continue to grow in size and brilliance. This is like the gigantic galactic fireworks display, that we had seen before, but now everything is in reverse. Each of the millions, or perhaps billions of stars, falls into their assigned positions, in obedience to their Creator's command. This is the creation of the pinwheel shaped galaxy, called the Milky Way. The entire creation of the galaxy of stars and planets appears to be a support system for planet Earth. We have just witnessed the "Dawning of Time."

The WORD looks upon His beautiful creation, and sees that it is good. Billions of stars sparkle like gemstones glistening in the velvet darkness of Space. It is a magnificent sight, and the Earth, with brilliance far exceeding that of all the billions of stars, is the centerpiece of His creation. Suddenly, He raises both arms and gives a thunderous roar toward the empty blackness of Space. Again, the mighty shout of the Creator Son of God rocks the corridors of heaven, and the terrified angels are startled once more. The WORD speaks the stars into existence before we are able to see them manifested. Far into the distance, thousands of fireballs begin to grow to tremendous size, then comes the familiar "big bang" that shakes the heavens. Billions of stars grow very large, as we watch them race to their appointed locations in the darkness of Space. Again and again, the WORD roars His mighty commands into the empty darkness, and the darkness flees away, and is replaced by a myriad of light and color. What a marvelous, galactic, fireworks display this is, and one that will never occur again. We have been privileged to observe the dark empty space of the "balloon" rapidly being filled with billions of stars. Each star is declaring the glory of their Creator God, to all who will see. This entire universe, as told in the Scriptures, was created for the glory of God, and for the pleasure of man.

Everywhere that the WORD pointed His arm and thundered His regal, majestic, commands, billions upon billions of stars were wonderfully created *"Ex Nihilo,"* out of nothingness. The Physical Realm is now filled in every part with a brilliant dazzling array of stellar gems. We have now seen the beginning of the Universe, or the "Space/Time Continuum," created by the WORD, for the glory of God, and so it is written in the Holy Scriptures:

> *"In the beginning was the Word, and the Word was with God, and the Word was God.*
>
> *The same was in the beginning with God.*
>
> *All things were made by him; and without him was not anything made that was made."*
>
> *(John 1:1-3)*

The sounds of the mighty shouts, and the big bang explosions, that have been echoing through heaven and the universe, appear to be subsiding at this time. The heavens and the earth have been created, as the plan of God continues to unfold before us. Like the angelic host who have been completely awe-struck by the power and might of their Creator, we too, have been overwhelmed by the forces that have been unleashed before us. We serve a powerful and awesome God. As we look out upon the vastness of space and see the sparkling display of God's creation, we are deeply stirred by the realization that all of creation was created by God's love, for our pleasure.

The angelic host appears to be recovering from their fearful and stunned condition. Everyone's eyes are riveted upon the WORD, as He slowly turns away from His beautiful creation, and begins to walk toward the Throne of God, passing through the honor guard. As the Son of God comes near, God the Father arises from the Throne to greet the WORD, and says loudly, *"Behold, this is my beloved Son, in whom I am well pleased."* The Temple angels begin to worship and sing praises to the Father, and to the Son, for the "great things they have done." The WORD stands before His Father, with arms outstretched and head bowed, He cries out in a loud voice, *"It is finished!"* As the glory of the Father intertwines with the glory of the Son, they are joined by the glory of the Spirit of God, who is always before the Throne of God. It is a thrilling, spectacular sight to see the Father, the WORD, and the Spirit, together as One God, in love and harmony. The Triune God of heaven has given us the record of what we have witnessed.

> *"For there are three that bear record in heaven, the Father, the Word, and the Holy Spirit; and these three are one."*
>
> *(1 John 5:7)*

All the angels of heaven joined in the celebration, for they could no longer restrain their emotions. The beautiful sight of the glory of the Triune God, against the backdrop of the newly created Universe, seemed to overcome the angelic host. All of their emotions were in play: fear and anxiety, reverence and honor, and love and joy all gripped their very beings. They have been witnesses to the awesome power and majesty of their Creator God, and so have we. There is a mixture of worship and praise, along with joyous shouting and jubilant conversation. This outpouring of song and elation could only be classified as a great heavenly celebration. We can hardly blame them for their exuberance, for we are rejoicing along with the entire angelic host. As we look more closely, there is one who has had very little demonstration of emotion, and has been conspicuously near the Throne, and yet appears always in deep thought. The archangel Day Star, has been apparently lost in thought, throughout the entire time of the creation of the Earth, and the Universe. Since we observed the creation from heaven, we have no idea how long it lasted, but how ever long the time frame was, Day Star appeared to be totally consumed by every activity. While every angel in heaven was focused on the WORD and His creation, Day Star was observing the Father, then the Spirit, and then the mighty angels about the Throne, as well as scrutinizing the whole host of angels. He seemed to take note of every emotion, every fear, every step of God's plan as it unfolded. He seemed to take particular pleasure in the creation of the planet Earth. His countenance appears to change, from being absorbed in deep thought, to elation and total awareness of the angelic celebration. He joins in the joyous activity before the Throne, and his radiant display is magnificent. His mantle encrusted with precious gems, swirls about, radiating a dazzling array of sparkling colors, like the reflections from a multi-faceted globe in a ballroom. His beauty and glory is greater than every angel of heaven, and is wonderful to behold. The Father, the WORD, and the Spirit are well pleased with the demonstration of praise and joy by the angelic host. It is a very happy time in heaven; every emotion seems to be pouring out. The celebration goes on for a long time, until the host is near exhaustion. As the heavenly praises begin to subside, the archangel Day Star raises his arms, as though to quiet the praises, and walks before the

Throne of God. The angelic host becomes hushed in total silence, and all the attention is upon Day Star as he begins to speak to his Creator God.

Day Star bows before the Throne saying, "May I speak my Father?" Out of the dazzling glory upon the Throne, comes the reply, *"You may speak, my son."* The great archangel replies with a very interesting request:

> "My Father, you have demonstrated your love toward us in a wonderful way. You have explained to us your plan for Creation and Redemption, and you created the angelic host to be a part in this plan. To show your love for us, we were privileged to witness the Creation of the Universe and the Earth. While the WORD was creating the Physical Realm, with its myriad of stars, I noticed that many of the angelic host were fearful of the Creation. It is my request Father, that you permit me to escort the entire angelic host into the Physical Realm to rid them of their fears. When you spoke to us about your plan of the Creation of Space, you assured us that, we would be able to move through Space, as easily as we do in the Spirit Realm, and that no harm would come to us. I would like to take them into the Physical Realm, to show them there is nothing to fear, and I promise to protect them, as though their God is with them. May I have your permission and blessing, my Father?"

The angelic host was quite surprised by the request by Day Star, and some even appeared to be slightly apprehensive about the prospect of soaring through the Physical Realm, while others seemed eager to explore the newly created Space. Excitement travels through the host very quickly. Their own individual apprehension seems to dissolve, and as a collective body, they began to look forward to the enjoyment of exploration in the new dimension of Space. The Father arises from the Throne, and an immediate hush falls. The Father replies to Day Star, and to the entire host,

> *"My beloved children, I am well pleased at the request of Day Star, who without hesitation, has accepted his role of protector and leader. He speaks truth when he recites My Words, because the heavens and the earth shall pass away, but My Words shall never pass away. All of the Universe shall be yours to pleasure in, and to explore, for there is nothing in the Physical Realm that can harm you. You are all my precious creation. I would not put you into harm's way; therefore I grant you permission to explore the Universe, for your pleasure and enjoyment. There is nothing unlawful for you in the Universe, accept touch not the glory of God upon planet Earth, obey my words, and heed my warning. This is the Universe and the Earth that the WORD has created, so rejoice and be glad in it."*

After God the Father finishes speaking, the Temple angels begin to sing praises to their wonderful, benevolent, Creator God and the angelic host joins in with great enthusiasm. Joyous singing rings throughout all of heaven. Their fears seem to fade, as the prospect of a great new adventure gripped their hearts, and they are eager to venture forth into the unknown.

III ADVENTURES IN SPACE

The angels are excited, as they look out upon the vastness of Space. There are billions of stars in every direction from heaven. Where would they explore first? What would they do in their first big adventure away from heaven? Suddenly, in the midst of the songs, praises, and cheers, Day Star arises from before the Throne, and slowly ascends above the host of angels. He is motionless for a long while, perhaps so that the host can see him in all of his magnificent glory, or perhaps he is gathering courage to enter the unknown. Nevertheless, Day Star is majestic to behold, for his radiant beauty is truly glorious. His mantle of precious stones dazzles in the glory of God's presence, and from his own glory that God has placed within him. As Day Star begins to slip out into Space, his countenance begins to change. His radiant countenance is no longer like when he was in heaven, but takes on a new appearance like the brilliant shimmering stars in the distance, and contrasts with the blackness surrounding him. He illuminates the whole area of Space with his glory, much like the WORD did, but his glory is nothing like that of the Son of God.

Day Star slips deeper into the darkness and then he pauses there, knowing that every eye in heaven is upon him. Still remaining motionless, he calls in a loud voice, for the angelic host to follow him. Cautiously, the angels venture into the darkness toward Day Star, the archangel. One after another, they set out into the unknown, until the blackness of Space is completely lit up with their glory. Their fearful silence gives way to joyful singing and rejoicing, as they began to playfully frolic in Space.

This scene immediately reminds us of the glorious advent in the night skies over the shepherds' field in Bethlehem that we witnessed, long ago. Only the Temple angels and the mighty angels guarding the Throne of God, remain behind, perhaps feeling it is a higher calling to be near to God. We watch them that followed him gather around Day Star, for instructions.

Day Star now speaks to the angels with a voice of authority and assurance. He tells them, "Break up into small groups, and just enjoy yourselves. After a reasonable amount of time, I will give a signal, and I want every angel to gather together, above the Earth. Remember that your God said; *'Do not touch My glory that is upon the Earth.'* Go now, and enjoy yourselves, and listen for my signal."

The small groups of angels bid each other good bye, as they slowly depart from one another in every direction. As they gain confidence, they begin to streak through the heavens at blinding speeds. Because their accumulated glories are quite brilliant in the darkness, the whole universe looks like it is filled with shooting stars. We can hear their shouts and singing from every part of the universe. Our curiosity prods us to join them, and get a closer look. Without a second thought, we soar out of heaven toward the nearest cluster of angels frolicking within a nearby galaxy. Our Eternity Explorer has no problem traveling between the Spirit Realm of heaven, and the Physical Realm of Space.

As we draw closer, we can observe some of the angels actually standing in a star, and are not consumed. We certainly are not going to attempt that because we are from the Physical Realm. Even in our Eternity Explorer, it seems prudent not to be reckless, and realize our limitations. We are able to hear new and different kinds of singing and rejoicing from the stars and even from other galaxies. The stars are singing together, perhaps, because of being in the presence of the sons of God, or they rejoice in knowing that they are an integral part of God's creation. We recall the Scriptures mentioning this very same occurrence, about the stars singing together, and the sons of God, or angels, were shouting for joy, a time of peace and harmony in the heavens.

> *"When the morning stars sang together, and all the sons of God shouted for joy?"*
>
> *(Job 38:7)*

The sounds are sounds of joy and innocence, like that of young children at a playground. The group of angels that we are with suddenly takes off toward another group at a distant galaxy. We follow them and are surprised at how quickly we arrive at our destination. The angels excitedly greet one another, and exchange stories of their adventures in Space. They realize that their fears are unfounded, and everything that God and Day Star had said was true. Without warning, there comes a very loud noise, like a thunderclap, that traverses the universe. The singing between the stars is interrupted by a signal from Day Star. Clusters of angels from every part of the universe begin to soar to the prearranged meeting place above planet Earth.

IV DAY STAR PLANS TO OVERTHROW GOD

We travel along with the angels, heading for Earth. As we look off into the distance, we can easily locate the Earth, with the glory of God radiating an enormous golden glow in the solar system. We spot the angelic clusters streaking toward this beautiful planet. As expected, the archangel Day Star is high above the radiant glory on planet Earth. His mantle of many colors shines brilliantly in Space, as it reflects the glory of God from the planet. Coming from every direction, angels circle about their magnificent leader, Day Star. Enjoying their adoration and how well they respond to his orders, with his arms outstretched, he begins to slowly turn and raise himself above the angelic host. A hush falls upon the excited host, until all eyes are upon him. Having their complete attention, he asks them in a loud voice, "Did you have fun?" They respond with shouts of joy and adulation, and even shouts of appreciation for leading them into the unknown Space. Day Star raises his arms once more, and an immediate silence falls at his commanding gesture. It appears that we are about to find out what he has been contemplating all along.

> "My fellow servants of God, did not God and I tell you the truth about Space? Did not God and I tell you the truth, when we said that you could exist in the Physical Realm, without being harmed? Did not the Creator God say, that we would live forever, and did I not prove it to you by asking God to let us venture into the unknown Space, under

> my care? Did I not prove my love and care for each and every one of you?"

The angelic host breaks into cheers for the archangel, Day Star, as he happily receives their praise and adulation. They, now, appear to have given all their trust to him. He asked many questions in his speech, but never waited for any answers. Their compliance was the answer for which he was waiting. Rallying around Day Star, the angels are intrigued by his words and cheer him, as though he is their idol. Awesome in beauty and great power and authority, Day Star is truly impressive. Elevating himself above the multitude everyone is moved by his presence. The archangel raises his arms for attention, and again a hush falls, as he once again begins to speak.

> "Our Creator has said that *'we shall live forever.'* God said that *'we cannot be destroyed.'* God said that *'you are strong and powerful.'* God said that *'we have a will, and the right to choose.'* God has called us, the mighty angelic host, *'His beloved children.'*
>
> Observe this beautiful planet below us. This is where God said that the WORD was going to create these puny mortals, called Man. God said, that they would be *'frail and fragile beings.'* God said, that they would have a *'short span of existence.'* God said that we would have to *'serve these faith-beings.'* Look at yourselves, you are a multitude of beautiful, magnificent, and powerful angels of God. God has said that you mighty angels shall be servants

to these puny faith-beings. Hear me my friends, I choose not to be their servant!

If I were God, I would not force my children to be servants. You should not be their servants, but instead they should serve you. You shall rule over these puny, inferior, mortal beings, as kings and gods. I shall be their king, and your king. I shall be their god, and your god. Together, we can rule the Universe. Together, we can overthrow God and the WORD, and I will rule, as your god from the very Throne of God. You must choose now, do you want to be a servant for eternity, or would you rather rule as a king and a god? What answer do the mighty angels have? Choose now, whom you will serve, do you serve Day Star, or do you serve God?"

The multitude of angels is spellbound at Day Star's speech. They seem stunned at first, but many quickly begin to shout cheers of approval. There is a chorus of dissenters, shouting their disapproval at the archangel's stand against God. The turmoil grows, as one by one each angel chooses in which side he believes. The angelic host seems to be divided; in whom they will place their loyalties and their trust. The situation is quickly becoming out of control. These heavenly titans are on the verge of warfare with one another. As Day Star looks on, he is very pleased with himself, and the resulting melee. Before the angry, shouting angels get out of hand, Day Star gives a loud shout, and the host quickly begins to restore order. He demands to speak further, and there is a deadly silence.

"We must return to heaven soon, and present our demands to the WORD, to create a race of mortals for us to rule over. If God refuses to comply with our plan, then we shall take the Throne by force. Either way, I will not be denied the Throne of God. I will be the King of the Universe. I will be as God. I will be God."

The mighty archangel slowly rises higher, as some of the angels loudly cheer their leader. Not wanting to miss any of their cheers, he waits for the adulation to completely subside, before signaling for the multitude to return to heaven. The entire angelic host slowly soars toward heaven, each one now has time to reevaluate his thoughts and position on Day Star's terrifying demands. Could we ever rebel against our Creator? Was Day Star really telling us the full truth? What will God do if He hears about this rebellion? There are many unanswered questions in the minds of the troubled angelic host, as they nervously return to the presence of their Holy God.

V SIN ENTERS THE HEAVENLIES

As we come nearer to heaven, along with the angelic host, we approach with great caution. This is a new experience for everyone to go to heaven from the Physical Realm. We have grown accustomed to the darkness of Space. The only light that we had was the light from the stars and the radiant glory from the angelic host. We were dazzled by the glory of God upon the planet Earth, but once we left the brilliance of God's radiant glory, we were in the blackness of the unknown. We are all engulfed by the darkness that surrounds us as we approach heaven. We seem to be traveling through a long dark tunnel, heading toward an incredibly bright light far off in the distance. The angels are deadly silent as we approach the heavenly portal between the Physical Realm and the Spirit Realm. We become bathed in the glory of God, and leave the dreaded darkness behind us. As the angelic host enters into heaven, and takes their place around the Throne of God, there is an eerie silence in heaven.

The returning angels probably expected welcome home cheers from the mighty angels around the Throne, and the singing of praises from the Temple angels. Instead, there is a deadly silence. The last of the angels arrive and take their places. The angels that remained in heaven notice something radically different about the countenance of about one third of the angelic host. We have taken note, also, that as the angels passed through the portal to enter heaven, the angelic glory of some angels intensified in radiance, while other angels lost their glory altogether. We also noticed the angels who had lost their radiance were not immediately aware of their loss. The last to enter through the portal of heaven is Day Star, who always seems to make a grand entrance.

VI WAR IN HEAVEN

The entire multitude gives out a gasp of utter astonishment, as the archangel slowly drifts toward the Throne of God and lands upon the glassy sea of crystal. He leers at his Creator and the WORD, who sits upon the Throne. Day Star quickly discovers there is something wrong: his countenance no longer radiates his beautiful glory. Turning away from the Throne, he rises above the glory of God. With a loud voice, he shouts commands to the host. "Arise, arise, we have been discovered," he yells to his followers, in a burst of panic. Before the angels can recover from this strange turn of events, Day Star seizes the moment. Now, realizing that he will never have a better opportunity to fulfill his great ambitions, he takes charge of the rebellion. Soaring over his followers, he shouts orders to them, "Attack! Attack! Attack!"

Above the din of the ensuing, furious battle, we can hear the roar of Day Star, shouting commands, "Attack the Throne," "Attack the Throne!" Deluded by his pride and arrogance, Day Star thinks he can overpower Almighty God. The mighty warrior angels about the Throne immediately encircle the Throne to protect their God. Angelic warfare is not like any war that was ever fought. Angels cannot die, but we can see that they can be damaged and hurt, because many have fallen in battle. Their power and strength is enormous, and the battle has taken on such fury, that we have been forced to move further away from the battle scene. The rebellious archangel, whose power is a terrible force to be reckoned with, has damaged a great many brave defenders of God.

The fury of the battle is indescribable, and the great roar of a rebellious multitude of angels is painful to hear. The harmony of heaven has been totally disrupted, and the foundations of the creation of God, seem to be threatened. The plans of God, for the Creation and Redemption of Man, appear to be in peril. As the battle rages on, we can now observe the

Triune God seated upon the Throne, apparently knowing the outcome must sadly and painfully await its conclusion. It must be heartbreaking, for Creator God to witness the rebellion of His own beloved creations. They were created in love, and are now rejecting that love in shameless open rebellion against their Creator. How sad that must make Him feel. We know that even as His creations rebel against Him, and reject Him, God still loves them.

The defenders of God have quickly devised a method of capturing the dark angels. Two angels would hold the arms of the rebelling angel, so that he could not damage anyone, nor could he fly away. The two angels would take their captive to the center of the encircled mighty angels, place him on the sea of glass, and hover overhead to prevent his escape. Thousands of the dark angels and the wounded are placed within this angelic protective dome. Since the rebellious angels no longer have their individual radiant glory, it is very easy to see who has chosen to be an enemy of God. The war in heaven appears to be coming to a conclusion. All of the wounded angels are gently brought into the circle of the mighty angels.

The archangel Day Star, is still battling the forces of God, when several of the mighty angels, led by the mighty warrior archangel Michael, lay hold of the rebellious leader. In a furious struggle, we hear the fiery rage that spews from the darkened archangel. Once he was the most beautiful creation of God, and now is screaming vicious, blasphemies at his Creator, but he finally realizes that his ambition to become as God, has perhaps been lost forever.

Michael, the warrior archangel, has subdued Day Star, and is bringing him through the dome of hovering angels, to the very same place before the Throne of God, where he was created. The shout of victory comes from the encircled mighty angels of God. The war in heaven is finally over. The hovering battle-wearied angels join in victorious jubilation. Michael has placed the rebellious leader in a kneeling position before God. His head is still bowed, perhaps postured to give the appearance of a contrite and repentant servant. His eerie silent, and completely motionless body, exhibits the awesome terror that one would have who is about to face judgment from a living Holy God.

VII THE TRIAL OF THE FALLEN ANGELS

The victorious cheers wane, as the Son of God rises from the Throne to address the angelic host. Slowly turning to observe every angel, there is great sadness in His eyes. Finally, His eyes settle upon the wounded within the ring of the mighty angels. With pity and loving compassion, the WORD raises His hands toward the injured angels and says *"Through the Love of the Father, be healed."* Wounded friends and enemies alike are restored to perfect health, as the onlooking angels gasp at the scene before them. With a majestic wave of His hands, the hovering angels are commanded to come to rest upon the glassy sea, as the WORD begins to speak to the host.

> *"My Beloved, It pains us to witness such horrifying tragic events. Harmony in heaven has been disrupted, and*

sin has entered the Universe. Sin is the rejection of God, and the disobedience to God's Word. You angels were deceived and believed the twisting of the truth of God. You have trusted in the creation, rather than the Creator. You have seen God, and you have seen Creation, and yet you did not believe and trust. You have sinned against yourselves, your fellow angels, and worst of all, you have sinned against your Creator God.

The Father has given all power and all authority to the Son. I am your Righteous Judge. This trial now begins, and my judgments shall be everlasting.

First, my judgment is upon the rebellious angels: You who are before me were created in love and grace, which was a gift from God. You have fallen from that love and grace, and from this point on you shall be forever known as "the fallen angels." You have fallen from the grace of God; there is no redemption for you.

Perhaps you have been wondering why you no longer possess a radiant glory? God is Love, and God is Light. When you were created in God's image, you were created with God's love and light within you. When you sinned against God, and rejected His love, the radiant glory of God, no longer existed within you. The love of God and His glory can never again dwell within you.

The wages of sin is death. The Creator brought you into existence by His word, and the Creator can deliver you

> *into non-existence by that same word. But God is merciful, and His mercy endures forever. Since sin cannot exist in the presence of a Holy and Righteous God, you shall all be banished from heaven forever. You shall receive your final sentence at the appointed time of the Great White Throne of Judgment, when all sin will be vanquished, and harmony will once again be restored in Heaven and the Universe.*

VIII THE TRIAL OF DAY STAR

The angelic host appears to be stunned at the severity of the sentence. Perhaps they now realize, how serious a matter it is to reject their loving, Holy, Creator God. The WORD, now seated at the right hand of God, points to the kneeling archangel, and with a loud voice cries, "Day Star, Arise!" The archangel slowly lifts his head, and with a deliberate and defiant pause, finally rises to his feet.

> *"Day Star, you are charged with the most serious sins of all. You thought that if you plotted to overthrow your God in a secret distant place, that God would not be aware of your treachery. The Holy Spirit of God is omni-present. His presence is from the deepest corner of Heaven to the farthest star in Space. Your God knows all things, there is nothing that is done in secret, or hidden from the Creator. Your name has been removed from the 'Book of Life,' and*

by your own words, you have sealed your own doom. Day Star, hear the charges:

- *You shall have no other gods before Me (Exodus 20:3)*

  You have proclaimed yourself to be a god, to reign over God's creations, as well as attempting to cause sedition and rebellion, to overthrow your Creator God.

- *You shall not take the name of God in vain. (Exodus 20:7)*

  You have accused God of not speaking truth before the angels. This is blasphemy, because God is Truth and it is impossible for Him to lie.

- *You shall not steal. (Exodus 20:15)*

  Day Star, by persuading these angels of God to abandon their first love and leading them into rebellion, you have stolen their chance to be in the presence of God, forever.

- *You shall not bear false witness. (Exodus 20:16)*

  You have testified falsely to your fellow angels about your Creator God, and theirs, for the purpose of lifting yourself to loftier heights before them. You are a liar, and you are the father of lies.

- *You shall not covet. (Exodus 20:17)*

  You have coveted the very Throne of God, and openly declared that you should be worshipped as

God. There can be no greater offense in the Kingdom of God."

The WORD finishes the long list of charges against Day Star, the archangel, as the entire angelic host awaits the sentence to be imposed. Would the Creator speak him into non-existence? What could the punishment be for such horrific crimes against the living, Holy God? If the punishment were to be unjust and vengeful, the angels that remained true to their God, and fought to protect their Creator, would now serve in fear instead of love. The Righteous Judge is about to give His verdict, as the entire host intently strains to hear every word. The drama is intense, as the WORD raises His right arm, and begins to speak in a loud voice for all to hear.

"Day Star, for your crimes against your God, your name shall no longer be Day Star, but from this moment on, you shall be called 'Satan,' which means Adversary and Accuser. Like your fallen followers, you shall not be destroyed, because of God's love and mercy.

Satan, hear this, the fallen angels may never return to heaven, but because of God's mercy, you may present yourself before your God at any time. You may present your objections to the Creator's plan of Creation and Redemption, and to be the 'Accuser of the Saints.' Know this Satan, I shall be their 'Advocate.'

I shall create a 'Lake of Fire,' and upon that appointed day of the Final Judgment, you Satan, and all those who follow you, shall be cast into it. All sin shall be removed from Heaven and the Universe, and cast into the Lake of Fire, and Satan and all his followers shall be tormented forever and ever.

Satan, you have said that you would be king of the Universe, but there is only one King of the Universe, and that is God. You are not a King, but you were a Prince in the Kingdom of God. Satan, you shall no longer be a 'Prince of the Light,' but from now on, you shall be known as the 'Prince of Darkness."

The WORD walks toward Satan and lays hold of him, as the archangel tries desperately to free himself. He quickly finds that he certainly was no match for the Son of God. The WORD begins to remove the beautiful bejeweled mantle that once lit up the entire Throne area with its radiant glory. The dazzling brilliance of this covering was magnificent to behold. It was a gift from the Creator, given in love to His greatest creation. The Son of God removes the beautiful covering from Satan, folds it neatly and carefully places it next to the Throne of God. The angelic host gasps with horror, to see the end result of sin. Satan's eyes now turn red with rage and hatred toward the Son of God. The beautiful body and glory, that once radiated the love of God, is now darkened. The countenance of Day Star once flooded all of heaven with beauty, now Satan's countenance is that of a twisted, hate-filled wretch. Seething, venomous rage now fills his very being, and even now, he appears to be scheming and plotting vengeance upon the Son of God.

IX SATAN CAST OUT OF HEAVEN

The WORD raises his right arm and points to the "fallen angels" before Him, and cries out in a loud voice. *"Michael, take hold of Satan, and cast him and his fallen angels out of heaven!"* Michael, the archangel, commands the mighty angels encircled about the Throne to lay hold of the fallen angels. Michael, himself, takes hold of Satan and slowly drifts above the glassy sea. The fallen angels are escorted slowly behind Michael and Satan, as they drift toward the portal into the Physical Realm, called Space. Each fallen angel looks longingly at the exquisite wonders of heaven for the last time. Each contemplates the foolishness of his decision to follow the cleaver deceit and lies of Satan. As the host reaches the edge of heaven, each of the fallen angels is cast out into the utter blackness of Space. The angels of God watch the fallen angels descend toward planet Earth. Since they no longer radiate with the glory of God, the fallen angels soon melt into the darkness. As they return, the angels of God thank God for the wisdom to resist the evil temptations of Satan.

As the angelic host gathers closely around the Throne, the Temple angels softly begin to sing praises to their Righteous God. We noticed the WORD has the mantle of Day Star upon His lap, and is plaintively looking at it. The mantle in the hands of the Son of God radiates its dazzling array of color throughout all of heaven.

The mantle or covering of the "Anointed Cherub" will be very important in the history of Israel. There is a fascinating connection between them, through the precious stones that are encrusted upon the covering. The Scriptures describe the mantle for us.

> *"....Thus saith the Lord God: Thou sealest up the sum, full of wisdom, and perfect in beauty.*
>
> *Thou hast been in Eden, the garden of God; every precious stone was thy covering, the sardius, topaz, and the diamond, the beryl, the onyx, and the jasper, the sapphire, the emerald, and the carbuncle, and gold; the workmanship of thy timbrels and of thy flutes was prepared in thee in the day that thou wast created.*
>
> *Thou art the anointed cherub that covereth, and I have set thee so; thou wast upon the holy mountain of God; thou hast walked up and down in the midst of the stones of fire.*
>
> *Thou wast perfect in thy ways from the day that thou wast created, till iniquity was found in thee.*
>
> *By the multitude of thy merchandise they have filled the midst of thee with violence, and thou hast sinned; therefore, I will cast thee as profane out of the mountain of God, and I will destroy thee, O covering cherub, from the midst of the stones of fire.*
>
> *Thine heart was lifted up because of thy beauty; thou hast corrupted thy wisdom by reason of thy brightness; I will cast thee to the ground, I will lay thee before kings, that they may behold thee."*
>
> *(Ezekiel 28:12-17)*

As we watch the fallen angels descend toward earth, let's discuss Satan's mantle. The nine precious stones, mentioned in the Scriptures, were part of the covering of Satan. In the numerics of the Bible, the number nine means "a gift from God," or (3) Trinity X (3) Heavenly Gift. The nine stones were taken from Satan and will be given to Israel, plus three additional stones. The twelve stones will be worn on the ephod of the high priest; each stone will represent one of the tribes of Israel. The priest's ephod, or breastplate, will represent the earthly gift from God. The numerical meaning of the number four, is "the Earth." Another way to express the meaning, would be, (3) Trinity X (4) Earthly = (12). The number twelve means "Government." God will deliver the twelve tribes of Israel from the bondage of Egypt. Under the Dispensation of the Law, God will give the tribes of Israel the gift of the twelve stones (Exodus 28:15-30). The tribes of Israel would then become the nation of Israel. It would be a nation that would be ruled by a Theocracy, or governed by God.

Under the Dispensation of Grace, or Church Age, the saints of God will be considered to be "Living Stones" (1 Peter 2:5), and they will be without number. The Light of God will radiate from within each believer. The Holy Spirit, who will dwell within each believer, will empower each "living stone," to project the light and love of God.

After the Millennial Age, will be the beginning of the "Eternal Dispensation." During this period, God will cleanse the earth with fire and make it holy (2 Peter 3:10-14). Once again, God's glory will be upon the whole earth. The WORD will create the future home of believers, called the "New Jerusalem." The sparkling walls of the New Jerusalem will be inset with twelve kinds of precious jewels (Revelation 21:18-21). God will take the twelve stones that were given to Israel, and He will keep eight of the original stones, and add four new stones for the foundation walls of the New Jerusalem. The number eight means "new beginnings," in the numerics of the Bible. As we know, the number four means "Earth," and the number twelve means "Government." When we put the whole interpretation together, the meaning is quite clear. It will be a "new beginning" for the "family of God," who will live in the "New Jerusalem," upon the "new earth," forever and ever.

As we return to the scene before us, we now see the fallen angels, as they near the earth. The angels of Satan are silhouetted against the radiant glory of God upon the earth. There is something prophetic about seeing the dark angels of Satan come between the glory of God on earth, and being in the presence of God in heaven. As the angels of Satan begin to land upon the planet, the beautiful glory of God disappears. We can now see the beautiful blue planet that the WORD created, before He placed the glory of God upon it. We must get a closer look at what Satan and the fallen angels are doing.

We are rapidly approaching the beautiful blue planet Earth, when suddenly we notice an enormous thick black cloud slowly cover the entire globe. We are not able to see water or land, but only darkness surrounds us. Finally, we penetrate the murky dense blackness that smothers us. We can barely distinguish any land features, because all of the land has been destroyed and made desolate. The dank, brackish waters are heavily polluted, and we cannot tell where the land ends and the waters begin. The earth is without form, and void; and darkness is upon the face of the deep. Does this mean that God's plan for the creation of man and his redemption, has now been thwarted? We must find out what happened to this beautiful blue planet. This is a great mystery to us. If we could find the fallen angels, perhaps we could find out how all this devastation took place. Let's see what we can learn about this horrible destruction.

Chapter 6

And God said, Let there be Light: and there was Light.

"In the beginning God created the heaven and the earth.

And the earth was without form, and void; and darkness was upon the face of the deep. And the Spirit of God moved upon the face of the waters."

(Genesis 1:1&2)

We have been cruising slowly through the vast dark cloud that presently envelops the earth. We have been observing the massive destruction of this once beautiful planet. From what we can see, there are only a few craggy outcroppings, and rubble strewn everywhere. The waters that once were so blue and pure, are now just murky swamps that look much the same as the dry land. This desolation is an abomination.

We search for any of the fallen angels to help us understand what has happened here. Now that our eyes have grown accustomed to the darkness we are able to see much better, although our vision is heavily restricted because of the thick dense cloud that engulfs us. Suddenly, we have come upon a small cluster of seraphim, who apparently are as startled to see us as we are to see them. Parking our Eternity Explorer we walk over to them. They are very curious about us, because they remember us soaring about in heaven, but thought we were just another one of God's creations. Well, that was true except that we were a creation of the future. We did not explain who or what we were, but set our goals to find out what happened to planet Earth. As we approach the seraphim, we are still very impressed with the size and power of these angelic creations. The leader of the group seems eager to talk with us, so all the others gather around us.

I THE FALLEN ANGELS

We ask the seraph, if the angels had any problems adapting to the climate and atmosphere of Earth. His reply was simply, "No!" We inquired further, if he would explain to us, everything that happened to them as they approached the planet. His eager reply flowed without hesitation.

> "As the host of angels came near to the this beautiful planet, we felt comforted by the glory of God that was upon it. The Glory reminded us of heaven and our Creator. We

thought that the exile to Earth was not harsh, because if God's glory was here, then we could know that God had not rejected us, even though we rejected God. But, as we entered into the brilliant glory cloud, it began to dissipate and disappear. When we landed upon the ground, we quickly gathered into groups to discuss what had happened.

We remembered the WORD had said, *'God can not be in the presence of sin.'* Because we have sinned against God, His glory could no longer remain on Earth. Sin separates us from God. We also realized that the glory of God upon the planet Earth was not meant for us, but for God's next creation, Man. We have brought sin into the Universe and now to planet Earth. Even though the glory of God was gone, the WORD had created a perfect pristine home for man. It was no longer holy, but it was perfect. The whole earth was like one beautiful lush garden."

The seraph became lost in plaintive thought and his memories. We quickly took advantage of the silence, and pressed the inquiry even further. We asked the seraphim, "Was there anything unusual that happened to the angelic host, once they arrived upon the planet?" Another seraph readily volunteered an answer to our inquiry.

"Once we arrived upon the planet, we soon learned that we had powers that we never used or needed in the Spirit Realm, or perhaps these powers are only unique to the Physical Realm. Since the angelic host is spirit, we have discovered that we can become invisible. In fact, invisibility has become our most natural state of existence in the Physical Realm. If we choose to be visible, then we can not see those who are invisible, but if we are invisible, we are able to see those who are invisible as well as those who are visible. I hope that makes sense to you. We found it quite an amusing phenomenon in the early times. We even played games amongst ourselves, using our invisibility, until Satan made us stop. We find it more comfortable to be in the invisible state, since we are spiritual and not physical beings. This is why you were unable to see any of the angelic host in your quest."

"This is all very fascinating, but you mentioned 'powers,' in the plural, what are the other things that you discovered since your arrival? We are very interested in anything that you would like to share with us." Another seraph from the back of the group boldly stepped up to us. Their apprehension seemed to readily dissolve as they noticed our obvious enthusiasm about their activities and well being. As he began to speak, the others drew closer to us.

"Some time ago, some of the angels found that they had the power to transform themselves into other forms, for a period of time. With some effort, I could transform myself into a being like you. We have not found a use for that power yet, so we discontinued doing it because the process expends a great deal of energy. We are also able to indwell our spirit being into others, but Satan said not to use this power unless he authorized its use. He must have some plan in mind for these powers, but he does not share many of his plans with the angelic host. He is quite secretive, and is his own private council.

Soon after the host arrived, we began to realize what the Creator meant by "time." We observed the countless rotations of the planet about the yellow star, and we discovered the changing of seasons and the passage of time. We discovered that all matter exists in the presence of time. We also recalled with trepidation, that the WORD said we would all be condemned, and cast into a 'Lake of Fire,' for eternity. Satan told us that 'it was not true, because a loving God would not destroy those He loved.' Satan also said, 'if God was telling us the truth about all sin being destroyed, he should have destroyed all of us then, since we were the only ones who ever sinned. God, therefore must be a liar. Just trust and believe in me, and no harm will come to you.' Most of the angels believed what Satan said, and the subject has not been discussed again."

"This is absolutely fascinating, and we truly appreciate your sharing with us. Could you tell us more about the time that you arrived on the planet? How did you live, was it difficult to survive in a new place, or in a new dimension, the Physical Realm? What was it like to live in a new and different realm?" Another seraph responded.

"When we left heaven, the WORD indicated to us that we could build homes and shelters, or even cities, if we chose to do so. We found it to be uncomfortable for a huge multitude to be massed in one location, so we agreed to divide into smaller groups. Since the whole planet was like one gigantic lush garden, it did not matter where we went, it was all so beautiful. We did erect homes and shelters for some personal comforts and privacy. It was quite pleasant, until one day, a group of the seraphim were gathered together in conversation. They were recalling the wondrous splendor of heaven, and how much they missed being there, and something horrible happened, that we would never forget.

Satan overheard the conversation and exploded with rage. He first verbally abused and cursed them, then without warning, viciously attacked every one of the seraphim. His mighty strength quickly overcame them, seriously damaging and wounding them. From that time on, every angel obeyed Satan out of fear. Satan's rage toward the WORD has grown deeper as the time of our freedom grows shorter, before the Day of Judgment. Satan has the power to influence the mind and will of anyone he is near, and has controlled the minds of most of our angelic host to be obedient to his every evil desire. It is only a matter of time until we, too, shall be hopelessly overcome by his evil vicious hatred. We have no spiritual or physical resistance to his will. We are doomed!

Since the terrible attack upon them, Satan has reduced all seraphim to the lowest rank of angels, and we have been

> treated like slaves and minions by all the other angels. Satan was the Anointed Cherub, so he has elevated the cherubim above the seraphim, and the few mighty angels, that believed his lies and were also cast out of heaven now have charge over the entire host. Absolute control over our entire being, is Satan's goal, and we can not resist. Satan controls the whole planet Earth."

We answer, "It is dreadful that such horrific powers have been unleashed upon this beautiful planet. We understand that sin spreads like a disease, until it grows and festers to its inevitable conclusion of death and destruction. Sin has its consequences, and everyone who sins must pay for his sins. Unfortunately, for the fallen angels, there will be no Savior, and there will be no salvation or redemption. God's plan of Creation and Redemption, is for man, not for the angels." The group of seraphim hangs their heads in silent acknowledgment of the truth in our statement. We continue, "None of the information you have given us, explains how or why this beautiful planet became so devastated. The beautiful garden planet you once enjoyed so much has been made a desolate, shapeless, swamp-like pile of rubble. What happened to the planet that once was created pristine and perfect, and possessed the glory of the Holy Almighty God upon its surface, but now suffers utter destruction? What kind of destructive powers were unleashed upon such beauty? What could possibly be the reason for such an abominable act? Who would dare to touch God's anointed?" One of the angels angrily snarled back at our impassioned questions.

"We had been enjoying the beautiful splendor of this planet for untold revolutions about the yellow star, until shortly after the incident of Satan's vicious attack upon the angels. Satan overheard one of his chief aides asking another; 'Do you think God will allow us to observe the creation of man?' Satan immediately went into a fit of rage beyond anything seen before. He lashed out at everyone around him, and the angels scattered everywhere. We could hear him cursing God, and his blasphemous railings against the WORD. His thunderous shouts could be heard around the planet. In the fury of uncontrolled rage, Satan began systematically destroying everything around him, including most of our dwellings. He commanded all of the angels to help destroy the beautiful planet Earth. He said that if the planet were in desolation and ruin, then Creator God would cancel His plans for creating His precious mortal 'faith-beings.' Satan's hatred for God has caused this destruction and desolation. He is our leader, and we must follow his every command. We cannot avoid him forever. Soon, even we shall succumb to his will and influence. I regret to say that one-day every one of us here, including you, will be his servants and minions. He is now searching the planet, like a roaring beast, seeking whom he may devour. When we see Satan, face to face, we shall be as he is."

"This has been the most informative and most terrifying revelation we have yet heard. It is obvious to us that we should never underestimate the evil powers of Satan, but in the same manner, we can take great comfort in knowing that the Son of God is so much greater. There is peace in your heart, when you have the WORD safely within to protect you from the wiles of the enemy of your soul. We want to thank all of you for your sharing this information with us. It is very helpful to understand everything that has happened here and we can sympathize with you about your plight. We can share something with you that may help. It is something that the WORD shared with the angelic host after you left heaven. He said, *'Resist Satan, and he will flee from you'* (James 4:7). You have been resisting him for a long time and that is a good thing, but without the love of God within you, it is an impossible task."

The seraphim, who were accustomed to the heavy dust cloud, looked up toward the heavens and recoiled in terror. We could not immediately see what had struck terror into these powerful seraphim. They were frozen in their positions, and searched for places to hide, but there were no places to be found. We still could not see anything because the eyesight of the seraphim is far superior to ours, but we were certain that there was something out there that brought terror to the hearts of the fallen angels. We strained to pierce the clouds, but it was difficult. Suddenly, we understood their terror. We saw a great light coming toward planet Earth, and we determined that it was not a meteor or a comet. We have seen this great light before. It was the brilliant glory of the WORD, the Righteous Judge, soaring toward us in majestic beauty. The Glory was like a thousand suns piercing the veil of darkness that entombed planet Earth.

II FIRST DAY: LET THERE BE LIGHT

The seraphim said they had thought that the Son of God was coming to cast them into the Lake of Fire, but we assured them that the Final Judgment would not occur at this time. God's plan was exactly as He said it would be; *"All sin will be vanquished in the fullness of time."* There will be at least another seven thousand revolutions of the Earth about the yellow star, called the sun, before the Great White Throne of Judgment. We calmed their fears by sharing with them, that it was our belief that the WORD was here not in the capacity of the Righteous Judge, but as the Creator God. We are about to witness something wonderful. The fallen angels stood in awe, as the darkness gave way to the glory of the Son of God.

The radiant brilliance of the glory of God that came down from heaven brought light into the Kingdom of Darkness on earth that was ruled by the Prince of Darkness. Every angel could see the WORD as He slowly soared above the Earth, and circled it several times. Every angel stared intently at the beautiful radiance above them, and was in awe of the Son of God. The WORD slowly came to a stop over an area not too far from where we had gathered with the seraphim. We invited the group of angels to go there with us, to see up close what the Son of God was about to do. We told them that it would be something wonderful, but they quickly declined. We could see the fear upon their faces, so we did not pursue the matter. We then returned to our Eternity Explorer, and sped to the position of the Glory light. Out of the Glory cloud, came a thunderous shout that was heard around the world. Our little craft jolted from the force of the sound, and the elements obeyed His command.

> *"And God said, Let there be light: and there was light.*
>
> *And God saw the light, that it was good; and God divided the light from the darkness.*
>
> *And God called the light Day, and the darkness he called Night. And the evening and the morning were the first day."*
>
> *(Genesis 1:3-5)*

At the command of God, the dust cloud that encircled the earth now disappeared and allowed the sun to shine once again upon earth. This is not the creation of the sun, but a recreation of the planet Earth. The WORD is restoring the earth that has been made desolate, being *"without form and void; and the darkness was upon the face of the deep"* (Genesis 1:2a). Meanwhile, the Scripture states that the Spirit of God was also participating in the restoration process, by cleansing the waters of the planet. *"And the Spirit of God moved upon the face of the waters"* (Genesis 1:2b). The Scriptures make it quite plain that the plan of Creation is from God the Father, and the execution of that plan is by the Son of God and Spirit of God. This is the first of many references of the Holy Spirit symbolized by water.

There is a theory existing today that states; when God said, *"Let there be light,"* He was commanding the whole earth to receive the knowledge of God, the Light that lights the souls of men. It is an interesting theory and is true, as far as it goes. The Light came into the world through Jesus, and everyone who believes on Him shall receive this Light (John 1:4-9).

And God said, Let there be Light: and there was Light.

This is a fundamental truth that is expressed in dozens of places in the Scriptures, but it does not seem to apply here at the recreation of the planet Earth. The only inhabitants at this time are the fallen angels who are fully aware that the Son of God is the Light, but they have rejected Him and cannot be redeemed.

The above Scripture also states, *"And the evening and the morning were the first day."* Perhaps, this is not a statement declaring that this was the first twenty-four hour day, or that the WORD started the world spinning on its axis. It may simply indicate that this was the conclusion of the first phase of recreation. Biblically, the term *"evening"* includes all of the dark hours of the day, from sundown to sunrise, or from 6 o'clock in the evening to 6 o'clock in the morning, with the dividing line at the terminator. The *"morning,"* includes all of the daylight hours, from sunrise to sundown, or from 6 o'clock in the morning to 6 o'clock in the evening. God uses this method of keeping time throughout the entire history of this world, from Creation to the Millennium. The Jews of today still follow this method, but the rest of the world has gone another way, other than what God has established. This, of course, is not the first time that man has discarded what God has established.

III SECOND DAY: THE WATER CANOPY ABOVE

With the sun shining so brightly, we could see the devastation that was wrought upon the earth. When the brightness of God's light shines upon the works of Satan, it becomes clear that the power of God must be unleashed to counteract his evil deeds. The seraphim that we were talking with have vanished from sight. In fact, we cannot see a single fallen angel anywhere. In the brightness of the light and in the presence of the Son of God, they have all become invisible, having sought a hiding place in which to cower. Even now, when the Son of God is alone, they dare not attack Him. We can observe the determination upon the face of the WORD, as He prepares to restore more of this hulk of a planet. Perhaps, we shall never know the pure and holy beauty that the earth once had in its pristine state. Maybe, we will see the exact duplicate, when God cleanses the earth with fire, and creates a New Heaven and a New Earth, after the Great White Throne of Judgment, and the New Jerusalem will be placed upon the Earth, like a crowning jewel. Silence befalls us, as the WORD is about to speak once again.

> *"And God said, Let there be a firmament in the midst of the waters, and let it divide the waters from the waters.*
>
> *And God made the firmament, and divided the waters which were under the firmament from the waters which were above the firmament: and it was so.*
>
> *And God called the firmament Heaven. And the evening and the morning were the second day."*
>
> *(Genesis 1:6-8)*

What a spectacular sight we are seeing here. At the command of the WORD, from the abundance of the seas that the Holy Spirit of God had cleansed and made pure in Genesis 1:2, water from the oceans rose up and multiplied as it soared higher and higher. The water rose from every part of the planet, and soon it looked like a gigantic balloon encasing the planet. As the waters soared higher, they continued to multiply much like stars multiplied as the WORD spoke them into existence. We also recall one of the miracles in the Scriptures, about the story of the incarnate Son of God multiplying the loaves of bread and the fish to feed five thousand people. It truly is a marvel to behold.

The water canopy is now thousands of miles above the surface, and we are beginning to see and feel the effects of this water barrier. The brilliant sunlight has been defused to a warm ideal condition, and with the warm gentle breezes blowing, it has become like the weather of a tropical paradise. The water canopy also blocks out all of the harmful "Ultraviolet Rays," that dramatically shorten life span. The canopy creates a "greenhouse effect," that will continually moisten all life upon the planet, once life is created. This is the reason that life forms will have extreme longevity of life. The Son of God has now completed His task of *"dividing the waters from the waters."*

"And God made the firmament," which is what we call the atmosphere, or the sky. The WORD created the atmosphere from the elements of the seas, and placed the atmosphere between the waters of the seas and the waters of the canopy, *"divided the waters which were under the firmament from the waters which were above the firmament."* This is the next step in the preparation of the earth to support life. The creation of breathable air, such as we know it, a combination of Nitrogen, Oxygen, Carbon Dioxide, and many other gasses, make up what we call atmosphere. It is important to note the last portion of the verse, *"and it was so."* Upon every command of the WORD, the elements were in obedience. Whether the WORD commands something to appear by way

of *"Ex Nihilo,"* creation out of nothing, or He commands things already created to change, all power is given unto Him (Matthew 28:18). However things are created, we can only stand in awe of such majesty and power of our Creator.

Twentieth century science cannot explain why every known star in the universe travels at the same speed, or why the stars are held in their positions. Science does not understand how molecules and atoms bond together, when the laws of chemistry and physics say that they should repel one another. The sciences of Geriatrics and Genetics cannot explain why people age and die. With the exception of the brain, every cell in the human body regenerates itself every seven years. What science does not understand, or is unwilling to accept, is that all things were created, exist, and are held together, or consist, by the will and command of the WORD; and He says so in His Word.

> *"For by him were all things created, that are in heaven, and that are in earth, visible and invisible, whether they be thrones, or dominions, or principalities, or powers—all things were created by him, and for him;*
>
> *And he is before all things, and by him all things consist."*
>
> *(Colossians 1:16&17)*

The preparation of the atmosphere is the fundamental necessity for all life. God calls the atmosphere by another name for a very logical reason: *"And God called the firmament Heaven."* Why is earth's environment called Heaven, and why is earth's air called the first heaven, when it was the last heaven created? The Scriptures were not written for God's benefit, but for man's. From God's point of view, Heaven, the home of God was created first. The second Heaven, Space, was created as the environment of the stars and planets. The third Heaven that was created was the atmosphere of earth. Although this was the order of creation, God reversed the order in the Scriptures for the benefit of man. As we stand upon planet Earth and look to the stars, we can comprehend our own atmosphere as the first Heaven, with the stars of outer space as the second Heaven, and the abode of God in the Spirit Realm, as being the third Heaven.

The conclusion of the WORD'S second phase of recreation is complete, and is punctuated by the now familiar phrase that concludes each recreative activity.

"And the evening and the morning were the second day"

IV THIRD DAY: LAND, SEA, AND PLANT LIFE

The WORD looked pleased with His work, as He went about the task of recreating the earth. It seems the Son of God takes pleasure in doing the will of the Father, in His role as Creator. We again draw close to the WORD, as He points toward the earth below.

> *"And God said, Let the waters under the heaven be gathered together unto one place, and let the dry land appear: and it was so.*
>
> *And God called the dry land Earth; and the gathering together of the waters called he Seas: and God saw that it was good."*

(Genesis 1:9&10)

At His command, the land rose above the waters, creating huge landmasses and islands. The great continents were formed, and the waters gathered together to form the oceans of the world. As we circled the earth with Him, we were in awe, as great mountain ranges rose to majestic heights. High craggy mountains of granite, limestone, and marble reached heavenward to touch the face of God. Lower mountains were formed that were rich in iron deposits, that proudly displayed their orange and brown hues in the warm sunlight. Pleasant rolling hills spread their beauty before the flat plain areas below them. Areas of swirling desert sands formed great sand dunes that revealed their undulating waves in the setting sunlight.

Islands arose out of the seas to declare their existence, as the WORD calmed the raging seas. We can see the great mountains already snow capped, because of their peaks plunging into the newly created atmosphere. Torrents of water flow from the mountains creating beautiful rivers, twisting through the foothills, seeking their passage to the seas. Great waterfalls and thunderous rivers, pour their waters into huge cavernous areas, and are filled to overflowing with water. These beautiful lakes are carved out by the hand of God, and seek to be a mirror to the heavens.

The Scriptures state: *"And God saw that it was good."* We were not close enough to hear the WORD make that statement, but we can understand its implication. The Son of God once created planet Earth pure and pristine, and when it was perfect, He placed the glory of God upon it and Earth was made holy. Satan destroyed the planet and the curse of sin came upon the earth. The planet is no longer holy and perfect, but in its recreation, can only be made *"good."* We can only imagine what this planet looked like when it was holy and perfect. Even if we did see it, there would be no adequate words to describe such wonder and beauty of God's holiness and perfection.

It was early morning and the sun was soon to peek over the horizon, when the WORD suddenly appeared from the east like the rising sun. The landscape lit up from His radiant glory. It was brighter than the mid-day sun, and yet we could feel only loving warmth emanating from Him. As He soared by, we could sense that something wonderful was about to occur on the dawning of this new day, and we wanted to be there when it happened. We pursued the Son of God, and we drew near just as He came to a hovering position, over the barren earth. He pointed toward the earth and commanded:

> *"And God said, Let the earth bring forth vegetation, the herb yielding seed, and the fruit tree yielding fruit after its kind, whose seed is in itself, upon the earth: and it was so.*
>
> *And the earth brought forth vegetation, and herb yielding seed after its kind, and tree yielding fruit, whose seed was in itself, after its kind: and God saw that it was good."*
>
> *(Genesis 1:11&12)*

At the creative command of the Son of God, all manners of trees, shrubs, and grasses, arose from the barren earth. The earth was quickly covered with greenery. As the morning sun broke over the horizon, we could see the sunlight dancing in the treetops. Some of the great mountains were still barren, and were majestic in their grandeur. The smaller mountain ranges had sprawling forests of hardwood trees, as well as coniferous trees. The forest floor was carpeted with undergrowth and shrubbery. There were, also, steamy, dense rain forests, where trees of many kinds and sizes, were crammed together, with the undergrowth struggling to glimpse at the sunlight. Every manner of tree, bush, and grass, had within itself, the seeds of reproduction. The seeds of the giant pinecones are far greater in size, than the tiny mustard seed, but all have the marvelous capacity of reproducing themselves. This is the first creation of life forms upon planet Earth. They are living entities, which have the cycles of life created within to reproduce at will. The WORD has, already created the water, air, and sunlight that are required to sustain life.

Mountain streams that wind their way down to the flatlands water the rolling hills. The babbling brooks and waterfalls fill the lush woods with their sounds. Beautiful glens and glades are wonderfully picturesque, and create idyllic landscapes that cover the once barren land. Pleasant meadows have a host of wild flowers and green grasses. There is color everywhere, with colorful fruit trees, flowering shrubs, wild flowers, contrasted by the backdrop of greenery, and crowned by the cloudless, pure light blue skies overhead.

As we soar past this green-carpeted landscape, we are now passing over the shifting sands of a desert which are very colorful in themselves, but are now accented by sagebrush, cacti, and tumbleweed that thrive in the hot sunlight. On we travel, in our imaginary Eternity Explorer, to the great expanse of the low lands, with its dried, brown grasses, sparsely scattered trees, and occasional patches of woods. Even the monotone-colored landscape has a unique beauty.

We have seen the marvelous patchwork of creation upon the face of the earth. There is an infinite variety of living plant life. It ranges from the giant trees of the forest, to the plants of the oceans, from the frozen tundra to the lush meadows and glades. The beauty that was created by the Son of God is difficult to describe, and should be experienced, to be fully appreciated. The infinite variety alone is not easy to grasp. Every species and every variety of plant life, is part of God's grand plan of Creation and Redemption. All of the atmosphere, the seas, and the plant life, serve a function and purpose in the great plan of the Creator God.

"And God saw that it was good.
And the evening and the morning were the third day."

V FOURTH DAY: SUN, MOON, AND STARS

The creative works of the WORD are magnificent to look upon. The early morning dew moistens all plant life, because there is no rain in this second ecological system. The greenhouse effect, created by the water canopy, keeps the whole earth well watered. The plant life thrives in this system, and is also protected from the harmful rays of the sun, by the water canopy. The plant life in the seas enjoys the luxurious sunlight causing photosynthesis, which produces a chemical change within the plants, because of the radiant energy of the sunlight. We are able to see the Son of God through the glory cloud, and it appears that He looks well pleased. We can't tell if His pleasure comes from His latest creation, or that He is nearer to His final and greatest creation. As we have seen before, the WORD is about to give His next creative command, and we are close enough to hear.

> *"And God said, Let there be lights in the firmament of the heaven to divide the day from the night; and let them be for signs, and for seasons, and for days, and years;*
>
> *And let them be for lights in the firmament of the heaven to give light upon the earth: and it was so.*
>
> *And God made two great lights; the greater light to rule the day, and the lesser light to rule the night: he made the stars also.*
>
> *And God set them in the firmament of the heaven to give light upon the earth,*
>
> *And to rule over the day and over the night, and to divide the light from the darkness: and God saw that it was good."*
>
> *Genesis 1:14-18)*

This is probably the most misunderstood section of this series of verses. Many believe that this is the creation of the sun, moon, and the stars. This is not the time frame of the creation of sun, moon, and stars, but it is the creation of "time." Genesis 1:14 describes the recreative adjustments upon the sun and moon that He had already created in Genesis 1:1. Let's start with the light that *"divides the night from the day."* This is obviously dealing with the sun and the earth. The WORD commanded the earth to rotate about an axis, so that the sunlight would warm and nourish every part of His recreated earth. This provides sunlight on a regular basis and each revolution is called a *"day."* Then the WORD created the exact elliptical orbit and tilt of the planet Earth about the sun to establish the *"seasons and years."*

Have you ever wondered how the earth began to rotate about its own axis? Considering the enormous amount of friction that the atmosphere has upon the earth's surface, why hasn't the daily rotation stopped? What or who sustains its constant revolution? The Bible tells us that all things *"exist and consist,"* or were created and held together, by the power of the Son of God (Colossians 1:15-17).

The *"signs and seasons"* refers to the changing of the seasons, when the leaves turn from green to brown, red, and yellow, as a *sign* of the new season. Some trees and plants shed their seeds at this time, to reproduce. Others reseed, when they begin their new cycle in the springtime. The sprouting of new leaves is a *sign* of new beginnings, with the hope of new life. Without the rotation of the earth, the planet would be frozen on one hemisphere, and baked by the relentless sun on the other. Without having the exact elliptical orbit about the sun, the planet would be unable to sustain life. The elliptical orbit does give this planet its *seasons*, and has been established by the WORD to maintain life by His command. The sun gives its light, warmth, and energy, through its own self sustaining power, and supports all life, according to God's plan. The star that the earth orbits is the sun that the Bible says is *"the greater light to rule the day."*

The second light in the firmament, or sky, is the moon. It is the *"lesser light to rule the night."* The moon radiates no light of its own, but it does reflect the light generated by the sun. It reflects this sunlight all of the time, except when the orbit of the moon takes it into the path of the shadow cast by the earth from the distant sun. The moon orbits about the earth on a regular twenty-eight days cycle, but the moon does not rotate about an axis.

The orbit of the moon was established by the WORD, to effect the earth in such a way, as to create seasons and times. The centrifugal force or pull of the moon upon the earth affects the tides of the seas and lakes. The moon controls many cycles and seasons. Planting seasons have been established by the phases of the moon. Many civilizations have used the phases of the moon to create calendars, or honored certain lunar days as holy days. The lunar phases have also been used as a guide for planting and harvesting. Many ancient civilizations have worshipped the great lights in the firmament, or have created pagan gods to represent these lights.

Let's look at verse seventeen, which tells us that God *"set"* the lights in the firmament. This is not an act of creation.

> *And God set them in the firmament of the heaven to give light upon the earth."*
>
> *(Genesis 1:17)*

He precisely *"set"* their pathways, or orbits, of both the sun and the moon, to make all of His other creations self-sustaining. There is no indication that these verses are talking about original creation. Verse sixteen is another example of the WORD establishing the sun, moon, and the stars to become visible, and *"set"* into motion in their orbits. The term *"set"* was used precisely. The Scripture speaks of how God made the greater and lesser lights to shine, and by the way, *"He made the stars also."* This casual reference to the stars is not the creation of the universe, but declaring that the stars were also made visible from earth, as part of the lighting system for the dark side of the terminator line. Again, the creation of the stars was accomplished in Genesis 1:1. The stars are an essential part of life upon planet Earth. Without the stars to guide them, sailors could never venture out into the seas at night. Since the moon traverses from horizon to horizon each night, there would not be any stable directional guide to follow, without the stars. Aside from their obvious usage, stars have another purpose, which is their awesome beauty that declares the wondrous glory of the Creator God, who created the heavens and the earth.

"And God saw that it was good.
And the evening and the morning were the fourth day."

VI FIFTH DAY: SEA CREATURES AND BIRDS CREATED

The green earth is rich and lush with all manner of vegetation. We can see woods and meadows awash with bright colors of fruit trees in blossom, and grand arrays of wildflowers, cascading over the rolling hills. The exquisite patchwork of color is everywhere. The grandeur of God's creation is wonderful. The great amount of species is enormous, and the infinite variety of variations of each species, is difficult to comprehend. Everything is amply watered by great numbers of streams and rivers, and by the constant supply of moisture in the pure fresh air. The water canopy feeds the atmosphere with water continually, and nourishes all plant life upon the earth. There is no rain, and the earth will not ever see rain until the worldwide flood at the time of Noah. The air is clean and pure, and the plant life is in great abundance in the seas and upon the land. It is a world of beauty and abundance, which has now been restored and is awaiting the next creative act of the WORD.

We can see the glory of the Son of God lighting up the skies over the horizon, and He is heading toward us. It is the dawning of a new day, and our excitement grows. The sheer excitement of seeing the glory of the WORD slipping over the horizon, and piercing the darkness, with a brilliance unknown in the galaxy, is awesome. The glen where we rested overnight is now glistening like a jewel in the approaching light. The morning dew shimmers in anticipation of the next creative act of the Son of God. As He soars overhead, we join Him, as we head toward the open seas. What a glorious sight, as we see the sun spread its palette of color across the horizon at dawn. The WORD points down toward the ocean and shouts His command at the seas, and then shouts toward the land.

> *"And God said, Let the waters bring forth abundantly the moving creature that hath life, and fowl that may fly above the earth in the open firmament of heaven.*
>
> *And God created great sea monsters, and every living creature that moveth, which the waters brought forth abundantly, after their kind, and every winged fowl after its kind: and God saw that it was good.*

And God said, Let there be Light: and there was Light.

And God blessed them, saying, Be fruitful, and multiply, and fill the waters in the seas, and let the fowl multiply on the earth."

(Genesis 1:20-22)

Before the WORD has finished His great command of creation, we can see the seas teeming with every kind of sea life imaginable. Everywhere we can see in the ocean, sea creatures are thrashing about at the surface, proclaiming a sort of praise toward their Creator. Great pods of leviathans, plesiosaurs or marine dinosaurs wave their long necks toward the Son of God, and huge schools of small fish thrash about in joy. Huge herds of the cetaceous group, such as whales, dolphins, and porpoises, breach the surface, reaching toward heaven and arching at the peak of their arc, and with a thunderous splash, return to the waters. It is a joyous celebration of life, much like the angelic praise toward their Creator, with of course, the obvious limitations of their physical powers, but it is obvious that there is no lack of enthusiasm.

We are now passing over the continental edge, and pausing over many lakes and streams. The waters are teeming with fish of every description, and all manners of crustaceans, and even microscopic water creatures. There are creatures of every color, size, and shape, and all of them appear to have joined in the celebration of praise to their Creator. Every creature that is created by the WORD has a designed purpose in life, everything from the minutest plankton in the seas, to the mightiest of the great leviathans, which we call dinosaurs. Every creature has a God-given, built-in instinct for survival in its environment. Each creature also has the capacity for reproduction of itself, after its own kind. The reproduction methods are as varied, as there are species, but in every case, the creation will procreate after its own kind.

While traveling slowly over the landmass, we encounter a new form of life. The air is filled with birds of every sort. They apparently saw their Creator approaching, and millions of birds began to circle about the Son of God. This seems very pleasing to the WORD, as He graciously stops in the midst of them, to receive and honor their adulation. As we look closely at the massive encircling flock, we can recognize many of the species, but most of them are not familiar to us.

The great soaring birds, like the condors, the majestic eagles, and many huge prehistoric pteradons, sail effortlessly from their lofty aeries high in the mountaintops. They join the sea-faring birds that followed us from the oceans to take part in the thankful celebration. There are enormous flocks of birds rising up from the forests, which are a magnificent display of brilliant color. Teeming with every color imaginable, the macaws, toucans, parrots, and hundreds of other species, parade their stunning plumage. There are birds that have tail feathers that trail behind them twenty-five feet or more, and radiate, shiny iridescent colors. What glorious sights these colorful birds are to behold. It is like watching a kaleidoscope that instantly produces a new myriad of colors, sizes, and shapes, at every turn of the hand. Quickly joining us, in this ever-increasing circle of celebration, are the flocks of birds from the woods and the meadows. There are many species of songbirds that join in with a chorus of praise to their Creator. Their songs are joyous, and appear to stand out above the raucous cries of the sea birds and the screeching of the soaring birds. As the day wears on, more and more birds rise to participate in the celebration of life. It has become a cacophony of sound, amid great splashes of color.

The sea birds are the first to leave, for their long flight back to the seas. Then the smaller birds that cannot endure long-lasting flights also begin to return to the glades and meadows. The long-legged marsh birds leave us, as well as the beautifully plumed and colorful forest birds. The cheerful songbirds stay as long as they can, but reluctantly return to the woods and meadows. One group after another returns to their own habitat. Only the great soaring birds remain, effortlessly sailing in the warm air currents. Finally, they, too, leave us for the lofty peaks of the majestic mountain ranges.

The WORD was well pleased with the display of love by His flying creations, but He begins to move earthward, as though He had some unfinished deed to accomplish. We do our very best to keep up with the WORD, as we travel to the lowlands. We can see great flocks of running birds and flightless birds, in this rugged land. It is now clear that the Son of God in all His great love and mercy has come to creations that could not celebrate with the birds of flight. This truly expresses His thoughtfulness and compassion to those that are different. We had forgotten about the flightless birds as part of His creation, but God is always mindful, even unto the least of these.

The flightless birds receive their Creator with great enthusiasm and excitement. There are many species of prehistoric dinosaur creatures that we don't recognize, as well as ostriches, moas, emus, kiwis, and a variety of ratites. They race in circles around the glory cloud of the Son of God who landed in their midst. Their raucous clamor is wonderful to hear.

They flap flightless wings, and kick up great dust clouds, as they race endlessly in circles, to show their appreciation to their Creator. This too is a celebration of life, and we are elated to witness such an event. The smaller birds began to tire, and other birds can celebrate no more. The WORD is well pleased with their show of love, as His beautiful radiant glory begins to rise above the excited flocks.

The sun is setting on this most exciting day. Magnificent splashes of color from this beautiful sunset wash over the rugged landscape. We remain here with the birds, as we all watch the glory of the WORD, disappear over the horizon, into the sunset. This is the perfect ending for a most exciting day. We have settled in this place to discuss the marvelous events of this day, and to anticipate the wonders of tomorrow. With God, it always seems that the best is yet to come.

"And the evening and the morning were the fifth day."

Chapter 7

And God said,
Let us make man in our image

The chirping and singing of birds in the pre-dawn light awaken us. We had exhausted ourselves late into the night by sharing the spectacular events of the previous day. Our camp is in a lush glen overlooking a valley, which is now blanketed with a thin morning mist. The campsite is damp from the dew that settled in the night and in the trees overhead, songbirds greet the new day with a joyous song. The sun is cresting over the horizon, splashing its brilliant, warm light upon tops of the nearby mountain peaks and birds in every direction are calling out their greeting to the morning sun. Sounds of life upon the planet give promise to a wonderful, new day that is filled with great expectations. A nearby stream rushes as it tumbles over a small cataract and spills noisily over a waterfall into the valley below. Fish splashing in the stream, songs of birds greeting us, and the rushing waters, are all wonderful sounds of life that are in great contrast to world that we first experienced.

The world that we saw before the recreation by the WORD was desolate and totally uninviting. God said, in the Scriptures, that *"the world was void and without form."* There does not seem to be a better way to describe that terrible sight. As we have seen, God is a God of creation and life, not of destruction. This new morning is a perfect example of the breathtaking sights and sounds of His recreation. We were anticipating to be witnesses to something wonderful that would be done by the WORD this morning, but we have not as yet seen Him. We have searched the skies from horizon to horizon for the brilliant radiance of His glory, but to no avail.

Morning sunlight engulfs the lush valley below, but we are concerned that the Son of God had forgotten us. He was always aware of our presence, and appeared to be pleased with our desire to continually be near Him. It has been our pleasure each morning to observe the WORD begin His recreative acts in our presence. It was as though He deliberately sought us out to observe these wonderful events. Not wanting to miss anything He is going to do, we decide to search for Him. Entering our travel craft, Eternity Explorer, we set out toward the rising sun in search of the Son of God.

Traveling toward the sun, we soon sight the magnificent, radiant glory cloud. Our craft slows to observe what is taking place before us. The WORD is making a zigzag course over the continent, going from one polar region to the other. He repeatedly pauses over an area, and then quickly moves on to the next spot, and pauses again. This pattern continues, as we try to figure out just what is occurring. We come to the conclusion, that we do not understand what is happening from our distant position so we move in closer as we have many times before.

I SIXTH DAY: LAND ANIMALS CREATED

As we move close to the Son of God, we can see huge groups of creatures springing into life on the earth below us. This is so fascinating; we pursue the magnificent glory cloud, and catch up with the WORD near the polar region. We can now observe closely the astonishing events before us. The WORD pauses, then points to the earth and speaks His command. Below, great herds of creatures appear: mammoths, reindeer, wolves, musk oxen, polar bears, and all manner of small animals. Each species is perfectly designed to adapt to the harsh weather of the elements of the tundra.

The WORD then turns back toward the warmer climates, but suddenly pauses over the high snow-capped mountain ranges, and again He points earthward and shouts His command. Upon a vast snow covered plateau, suddenly appeared herds of elk and moose, and herds of fleet footed sheep and mountain goats appear. Packs of foxes dark colored wolves, grizzly and kodiak bears, and a large group of sleek, saber-toothed tigers pace excitedly close by. Thousands upon thousands of little creatures are scurrying about the milling herds of large land animals. The newly created animals, in awe and respect, silently gaze toward their Creator. It is a wonderful sight to behold, as the creations observe a reverence toward their Creator.

The WORD acknowledges His new creations, and we move toward the more temperate areas. We have finally discovered why this creative process is so different than all the rest. The Creator is going from polar region to polar region, like a farmer walking from one end of his field to the other, "a sower casting his seeds" to the ground as he goes. The Creator is planting, or seeding, His creations in the climate and environment, in which each species is best suited for its survival. The sea creatures can migrate to any part of the world they choose, for warmer, or colder, or deeper waters, or perhaps, motivated by preferred or seasonal food supplies. Like the sea creatures, birds, too, are able to migrate to any part of the world at will. Many species of birds fly thousands of miles every year for warmer weather, and then return again. Birds and sea creatures have the capacity to migrate great distances. Land animals, on the other hand, are mostly locked into their own habitat that is controlled by climate and food supply.

We have arrived at a warmer climate, where the woods and foliage are lush, with an abundance of lakes, streams and ponds. The WORD once again pauses, points to the earth, and shouts His command for the earth to bring forth living creatures. Herds of huge beasts, animals that we know as dinosaurs appear in the large meadow below. Dinosaur in Latin means "terrible lizard," but we could not recognize most of these strange creatures. The ones that we knew about were like old friends to us. The huge brontosaurs are the first ones we spot, because of their enormous size. Brontosaurus in Latin means "thunder lizard," a good name for a creature that shakes the ground when it walks. There are the fierce looking tyrannosaurs, which means "tyrant lizard," with their huge heads and powerful jaws. Grazing in the colorful fields of wildflowers were the herbivores, the stegosaurs and the triceratops. The stegosaurus, with its armor-plated spine and spiked tail is aptly named, its name means, "roofed lizard." Along with them is my personal favorite, the triceratops, which means "three horns on the face." These huge animals are marvelous to watch, but the meadow quickly fills to capacity. These great animals, all are gazing at their Creator, as the WORD moves on to another area similar to the last one, except this area has a beautiful waterfall nearby. This time we want to get close to the Creator to hear what His words are, as He points to the ground, we hear His words:

> *"And God said, Let the earth bring forth the living creature after its kind, cattle, and creeping thing, and beast of the earth after its kind: and it was so.*
>
> *And God made the beast of the earth after its kind, and cattle after their kind, and every thing that crept upon the earth after its kind: and God saw that it was good."*
>
> *(Genesis 1:24&25)*

Appearing in the meadows below us, are totally different types of creatures, some that might be considered as farm animals, others are designed for the woods and streams, such as beavers, deer, squirrels, raccoons, and all manner of rodents. There is something we are missing at this height, so we settle to the ground for a closer look. What we had overlooked before was a whole world of very small creatures. The ground appeared to be covered with millions of bugs and insects. There are thousands of species, of every conceivable shape and color. There are rodents, snakes, and lizards, and every kind of reptile, that could burrow, climb, slither, or jump. The scope of the variety here before us was beyond knowing.

We move on to the equatorial plains area with its vast open ranges and occasional wooded patches and watering holes. In the distance we can see the green mountains rising off the flat plains of the veldt, parched, brown grasses contrasted and intermingled with lush green grasses and shrubs. The WORD pauses at the plains expanse, and again command: *"Let the earth to bring forth the living creature after its kind."* Suddenly appearing below us were great herds of buffaloes, zebras, wildebeests, elephants, and giraffes. There are huge groups of gazelle-like animals of every size, color, and stripes, bounding and prancing with excitement. Near the river, there are herds of hippos and crocodiles heading for the cool water and off in the tall grass, we can see large families of the big cats, lions, leopards, and cheetahs. The veldt is quickly being filled up with an enormous variety of land animals, many of which are quite unfamiliar to us. Our excitement is almost uncontainable.

Overhead, we notice that some of the great soaring birds have found this to be home. Flocks of circling condors and vultures accompany the pteradons, with their twenty-five foot wingspans. Great flocks of brilliantly colored macaws and parrots are now arriving; knowing instinctively that the jungle would be home to them. They are keenly interested in the miraculous sight below. The great herds are all facing their Creator, as the elephants lead their praise with their trunks raised, trumpeting loudly, while the other herds join in the celebration. The okapis and the kudus joined in, and the bleats of the impalas and gazelles can be heard, as well. The neighs of the zebra, and the snorts from the buffalo herds, seem to be overcome by the deep roars that come from the direction of the big cats. Even the circling giant birds overhead took part in the celebration, with their unmistakable screeching.

The WORD quickly heads for distant mountain ranges, that we had seen earlier, and we follow as fast as we can. We are at the edge of the savanna, where the grasses are lush and the mountainous, dense jungle seem to be endless. The tropical heat is a welcomed change from the arctic climate that we experienced earlier. Once more, the Creator points toward the earth, and shouts His command: *"Let the earth bring forth the living creature after its kind."* In the clearing below, great herds of creatures instantly appear. The noise of the snorting, snarling, and screaming animals is extremely loud. The din mostly comes from the great families of primates: the gorillas, chimpanzees, and a host of species of screaming monkeys. There are giant sloths, rhinos, tigers, ocelots, and an enormous variety of small scurrying creatures. We can see an infinite variety of reptiles, snakes, and insects that would thrive in a jungle environment. As usual, there are hundreds of species that we cannot identify, but we know that each of these species was totally adaptable to these surroundings. Each species is not only specially designed to survive in this environment, but is designed to reproduce its own kind. The grand design of all the creations that we have been privileged to witness, can only be the plan of the Great Designer.

The screams and chattering of the primates, and the roars of the big cats, seem to subside, as the brilliant radiant glory of the Creator descends closer to the vast herd of creatures. They too, express their joy and elation toward their Creator. Our excitement grows to a new height this day, as we are privileged to observe the acts of creation by the Son of God, and to see Scriptures fulfilled before our eyes. We are eager to see more of God's marvels, but the WORD motioned for us not to follow Him. Although we are puzzled, we are, of course, obedient to His will. We are probably just slowing Him down. His radiant glory streaks toward the ocean, perhaps to create the multitude of animals that are unique to the hundreds of islands, and to the island continent.

We decide to rest here and marvel at the vast array of creatures that are before us. In the Scriptures, the term used for *"creatures"* is the Hebrew word *nephesh*, usually translated as *"soul."* The term soul implies conscious life, as opposed to plant life, which has unconscious life. These wonderful creatures that are before us, surely have a conscious awareness of themselves and of others around them. The term *nephesh* or *soul*, as used in association with man, takes on a totally different meaning. The soul of man is eternal and not only has a life consciousness, but has God-consciousness. We will have to continue this discussion later, because we can see the radiant glory of the Son of God, coming over the horizon toward us. It is like a rapid sunrise coming from the west, and very spectacular to observe. It is an eerie feeling to know that the Creator of all things is taking particular interest in us. While He is maintaining an obviously busy schedule, He is showing great compassion toward us.

THE CREATION OF MAN

As the radiant glory cloud comes closer, it appears to be so much more brilliant than before. Either the elation of the WORD has made Him more radiant, or God the Father has joined with the Son of God and the Spirit of God for the greatest of all His creations: man. We have been summoned to follow by the Creator, and our excitement grows beyond our control. With great anticipation, we try desperately to catch up with the divine fireball streaking through the second heaven. The Creator slows and begins to descend, over a lush land area where two major rivers join together. We land in a meadow with a patchwork of wildflowers near a grove of trees that are filled with songbirds welcoming us with their joyous songs. Beautiful butterflies dance from flower to flower. There are squirrels and chipmunks chasing each other in an endless game of frolic. A family of deer cautiously peers out from the edge of the woods to see what is causing the excitement. All of the creatures seem to be quite calm in the presence of this awesome sight of the glory cloud of the Creator. Their calmness is in sharp contrast to our excitement. As we had expected, the Creator is about to speak, and we hear the clear, soft voice come from within the glory cloud.

> *"And God said, Let us make man in our image, after our likeness; and let them have dominion over the fish of the sea, and over the fowl of the air, and over the cattle, and over all the earth, and every creeping thing that creepeth upon the earth.*

> *So God created man in his own image, in the image of God created he him; male and female created he them."*
>
> *(Genesis 1:26&27)*

The Son of God has proclaimed a very interesting statement. The confirmation of a "Triune God" is declared in a straightforward manner. *"Let us make man in our image, after our likeness."* is quite clear about the plurality of God. The terms, *"us, our, and our"* are all plural, when God is speaking about Himself. The next interesting statement is that man was created with the capability to reproduce both male and female offspring. Aside from the newly created animal and plant life, there are no female species anywhere in heaven or in the universe. The Creator God is masculine and the angelic host is male. The female of the gender does not exist anywhere in the universe. The female counterpart of man has not yet been created, but God boldly proclaims *"...and let them have dominion"* as though it were so. Furthermore, through the creation of man, God is preparing for Himself, a "Bride."

The WORD reaches down and scoops up handfuls of dirt, and begins to lovingly mold the dust of the earth into a shape that is pleasing to Him. It seems much like the potter molding his clay into the exact perfect shape that he desires. The Creator does not speak man into existence, as He did in creating everything else in heaven and in the universe, but is personally forming man with His own hands. Everything that exists was spoken into existence by the Creator God, with the exception of man. The creation of man is obviously being accomplished with great love and tenderness that is far beyond our comprehension.

The radiant glory of the Creator is difficult to see through, so that we cannot observe exactly what is happening. The entire glen is ablaze with His divine glory and, yet, none of the creatures flee, because they all seem to realize that there is nothing to fear in the presence of their Creator. As we take a closer look around us, we observe the creatures in the area are solemnly watching this marvelous occurrence at a respectful distance. Suddenly, something wonderful happens within the glory cloud. Man becomes a living being.

> *"And the Lord God formed man from the dust of the ground, and breathed into his nostrils the breath of life; and man became a living soul."*
>
> *(Genesis 2:7)*

The glory cloud of God rises slowly above the glen, enabling us to see His creation of love. Man, lying on the very ground that he was created from, is alive and breathing but appears to be in a deep sleep. The brilliance of His glory seems to increase, as the WORD rises higher. It is my guess that God the Father joined God the Son, and Spirit of God, for this miraculous event. The increased radiance is probably the result of love and pleasure derived from the creation of man. Unexpectedly, the Creator God slowly soars off until He is out of sight. We are apparently left to watch over God's creation.

> *"And God saw every thing that he had made and, behold, it was very good..."*
>
> *(Genesis 1:31)*

When the brilliant presence of the Creator was here, we did not realize that the day was ending. In the waning daylight, we notice that the man also has radiance emanating from him. It is as though the "signature of God" is upon him, and that this creation is pure and holy, belonging to the Triune God. While witnessing the spectacular events of the last few days, we had completely forgotten about the fallen angels and the evil forces that are upon the planet. The glory of God that is upon man, would be warning enough for the fallen angels, but if Satan knew that God's greatest creation lay helpless before us, we could all be in danger.

We are in awe of the beauty of man. Each feature is flawlessly sculpted to perfection. Every line and every curve flows in perfect symmetry. Man was designed by the Great Designer to be pure, and holy, and to live forever. When God breathed life into the lifeless body, He was breathing His own Holy Spirit into him, which gave man eternal life. Without the Spirit of God in man, he is still only the lifeless dust of the earth. Man has now become a living soul, made in the image of God, designed to be an eternal Son of God. Man is now being covered by the lengthening shadows of the sun setting in the western horizon. Many of the creatures draw closer to his still figure lying in the shadows, and yet he continues to glow with a golden aura from the touch of his Creator God.

"And the evening and the morning were the sixth day."

II SEVENTH DAY: THE SABBATH DAY CREATED

The next morning, we were awakened by the joyous songs of the songbirds, which were greeting the first slivers of sunlight cresting over the distant hilltops. As the glen quickly became bathed in the golden morning sunlight, the little creatures began to scurry about in their playful manner. The birds were chirping loudly, as though they were the appointed wake-up system for the neighboring residents. The grasses and the trees are moist with the morning dew. The droplets of dew sparkle in the sunlight like tiny jewels. They shimmer and glisten as the gentle breezes move through the glen. The deer are particularly curious about the man that is still lying before us in a very deep sleep. Cautiously, they approach the new creature in their woods, and even nudge him with their noses, but there is no response.

We are becoming anxious to see the WORD, and begin to search for His telltale glow in the sky. The nearby brook, with its continuous lulling sounds calms our anxieties, but we continue the search. It doesn't seem likely that the Creator has forgotten about us, but we are concerned just the same. We begin to recollect all of the wonderful events of these many days when we realize that this is the seventh day since the Creator came to restore the earth, and create life upon it. It is a good possibility that we would not see the WORD today. To pass the time, we discuss the miraculous events of the week. As we talked about the seventh day, we found that it was as interesting as the six creative days.

The day that we are now experiencing is the first time that God has honored any day. God declares the seventh day of the week not only to be a day of rest, but He blesses it and He sanctifies it. The seventh day of "rest" in the Hebrew is called the "Shabbath." The Scriptures are quite clear as to why this day is to be honored above all other days of the week.

> *"Thus the heavens and the earth were finished, and all the host of them.*

And on the seventh day God ended his work which he had made; and he rested on the seventh day from all his work which he had made.

And God blessed the seventh day, and sanctified it, because that in it he had rested from all his work which God created and made."

(Genesis 2:1-3)

The "Sabbath" was not created because God needed a rest, but the Sabbath was created for the benefit of His greatest creation, man. It is man who will receive the blessings for honoring the day that was sanctified, or made holy. The day of rest is just that, man will need a day to set aside to physically restore his energy, but more importantly, it is a time to give honor and glory, and worship, to man's Creator.

The number "seven" is God's number of "Completeness," and "Spiritual Perfection." God uses the number "seven," exhaustively, throughout the Scriptures. The following are just a few examples of how the number seven has effected even the history of man.

- (Genesis 2:2-3) On the seventh day of Creation - <u>God Rested</u>
- (Exodus 16:23) On the seventh day of the week - <u>Man Rested</u>
- (Leviticus 25:2-5) Every seventh year, the land, the orchards, and vineyards were unworked, the Land Sabbath - <u>The Land Rested</u>
- (Leviticus 25:8-12) On the year after the seventh "Sabbath Year," or (7 X 7 years), Israel had to declare a Jubilee. Every fiftieth year all debts were canceled, and all slaves were freed - <u>Israel Rested</u>
- (Revelation 20:6) In the seventh millennium from the creation of Adam, Jesus will rule the world for 1000 years - <u>The World Rests</u>

There was no mention of the Sabbath, from the creation of man to the giving of the Law to Moses on Mount Sinai. As part of the Ten Commandments, the Sabbath was a continuous reminder to Israel, for a time of rest and meditation of her special relationship to God. The Ten Commandments were given to Israel, under the Dispensation of the Law, to make them aware that they were sinners and needed a Savior. Since no one could ever live up to the Law, God set forth a standard that was only fulfilled in His Son. The first four of the Commandments pertain to their relationship with God, and the next six pertain to Israel's relationship with their fellow man. In the Ten Commandments, God reiterates His desire to keep the Sabbath a holy day and to remember their Creator.

> *"Remember the sabbath day, to keep it holy.*
>
> *For in six days the LORD made heaven and earth, the sea, and all that in them is, and rested the seventh day; wherefore, the LORD blessed the sabbath day, and hallowed it."*
>
> *(Exodus 20:8,11)*

In the New Testament, Jesus made it crystal clear what the Sabbath day was all about. He chastised the Pharisees for being so legalistic concerning the Sabbath, and misinterpreting God's intent to the people about the Sabbath.

> *"And he said unto them, The sabbath was made for man, not man for the sabbath.*
>
> *Therefore, the Son of man is LORD also of the sabbath."*
>
> *(Mark 2:27&28)*

The Sabbath, or the seventh day of the week, has been honored by the Jewish people, ever since Moses received the Law on Mount Sinai. Christians however, also observe a day of rest, not on the last day of the week, Saturday, but on Sunday, the first day of the week. The Apostles and disciples honored the first day of the week, because it was the day that Jesus Christ, the Son of God, rose from the dead. Christians have been honoring the first day of the week ever since the early church, to honor the Day of Resurrection of the Savior. It is interesting to note that the last day, the Sabbath, commemorates the completion of *Creation*, by the Son of God, while the first day of the week commemorates the completion of *Redemption*, by the Son of God. The plan of Almighty God, *for Creation and Redemption*, is honored in these two observances, the Sabbath and the Lord's Day.

III THE GARDEN OF EDEN

We have spent most of the day resting in this beautiful glen, and discussing all the wonderful things that God has done. The man lying beside us is at rest as well. This, perhaps, is exactly what God had in mind, when He created the Sabbath. It was an exhausting week for us and we certainly needed the rest, but it was refreshing to our spirits to contemplate and discuss the miraculous works of the Creator. It is truly a time of refreshing for the spiritual part of man, as well as for physical regeneration.

The evening shadows are lengthening about us, as we gather some of the fruits and berries, which are in abundance. We are thankful for the nourishment that God has generously provided. As we savor our delicious repast, we continue our discussion. Looking upon the man lying before us, we have a question. Is the WORD through with His creative works, or is the creation of man just the last of the creative groupings, that are known as the six days of creation? What about man's home in the garden, and what about man's mate? We recalled that the end of the "sixth day of creation," ended in chapter one. The creation of the garden and the mate of man are all described in chapter two. Our scenario, as it is unfolding before us, is in the sequence that is in the Scriptures. We have witnessed the creation of man at the end of the sixth day, and we are now experiencing the seventh day, the Sabbath, and we have not yet seen any other creative acts on these matters. Tomorrow may bring new and glorious creations by the WORD. Our excitement soars at the possibility of witnessing more of God's plan of creation unfolding before our eyes. The night is upon us and we once again gaze upon the radiant glow of man close by us. He has given us assurance of God's love, and gentle peace come over us, as we become lost in our dreams of the wonders that are before us.

The chirping and trills of the songbirds awaken us in the early morning hours, singing glorious songs of praise. Peeking over the eastern hillside, the sun floods the glen with brilliant shafts of light. With the dawning of a new day, our anticipation mounts to new heights, for we knew that we would witness marvelous events this day. This is the eighth day since the beginning of the recreation of the earth. The number eight in Biblical numerics means *"new beginnings."* Even the nearby brook seems to add its voice to the excitement, with the roar of its resounding rushing waters. All of the creatures of the woods seem to expect something wonderful is about to happen, as each one cautiously edges closer to us and to the still, glowing figure lying near by. We are scanning the heavens for the glorious appearing of our Great God and Creator.

Suddenly, the gentle cooing ceases from the morning doves in the trees above us. Deer with their keen senses alerted, turn to gaze off into the distance. Squirrels seem to sense something, as well, as they stopped their scurrying, in their endless game of chasing one another. Songbirds who are no longer singing their usual joyous greetings to the new day fall silent. The eerie silence is a new experience in the glen. No sound of any kind can be heard, with the exception of the lilting, swirling sounds of the nearby brook.

Intense excitement grows, as the WORD appears over the horizon, and lights up the already bright morning skies. The Creator hovers over us momentarily, then gently descends toward the man below. All the creatures scurry away from the man, and observe from a respectful distance. The Creator lovingly cradles His most beloved creation to His bosom. Man is lost in the arms of God, for His glory completely covers the new creation. Man is in oneness with his Creator, and the Creator loves His creation. Slowly they ascend above the glen, giving us an opportunity to race to the Eternity Explorer.

We follow as closely as we can, but we do not have to travel very far, because we can see the unmistakable golden glow of God's glory upon this tract of land between the two great rivers, and bordered on the other two sides by two lesser rivers. There is no doubt that this is our destination. From our vantagepoint above the area, we can see that God has marked His territory upon this planet, proclaiming to all other inhabitants of the earth not to trespass, for this ground is holy ground. Slowly, we begin to descend toward the golden glow. The WORD is not in a hurry, perhaps wanting to savor every moment of this wonderful advent. We pierce the golden veil, and are startled to find that we have entered into some sort of dream world. We have entered into the Garden of God.

"And the LORD GOD planted a garden eastward in Eden; and there he put the man whom he had formed."

(Genesis 2:8)

We have arrived at the incredible home of man, a place that the Lord God Almighty has prepared for His greatest creation. This is the beautiful Garden of Eden, the first home of the first man. The Garden is spectacular, almost beyond description. It is like nothing that we have ever before seen. The beauty here is breathtaking and challenges the human senses to comprehend its uniqueness.

Once we have penetrated the golden glory of Eden, we no longer notice the aura of God's glory, for we are now engulfed in that glory. The air is pure and sweet to breathe. The wonderful aroma that permeates the air is exciting to our very souls. We notice exotic flowers growing everywhere, some are known to us, but most were never before seen by mankind. Flowers with deep rich colors, dramatically stand out, while others have delicate, soft, pastel tints. There are even flowers that are translucent and some are transparent, that exude not only exotic scents, but soft tinkling sounds emanate from them like a distant chorus of wind chimes.

Trees of every kind are here. Some bare nuts, while others bare all manner of fruit. Some tall, sturdy trees are host to climbing vines interwoven through their upper branches to display their beautiful array of flowers. Gentle, warm breezes brush our faces and we can detect new and different aromas that are wafting from a distant part of the Garden of God. The exotic sights, sounds, and aromas are thrilling to experience, and we have only been here a few moments. This is a place that one could happily spend a lifetime exploring its unmatched, pristine beauty.

We head for an inviting grove of trees, with a small, trickling stream leading to a placid lagoon. In the lagoon is a family of brontosaurs feasting on the succulent underwater plants, while, on shore, carnivores and herbivores are quietly grazing the lush grasses. As we approach the grove, several peacocks greet us with full plumage extended and trumpet their shrill cry to announce our presence. In the tree branches above, warblers and meadowlarks join in the chorus of greetings. Soon after the unsolicited announcement of our arrival, animals of every kind seem to be heading toward us. Were they approaching because of our presence, or was their interest in something else? In our excitement about experiencing the wonders and beauty of Eden, we have momentarily forgotten that this is not about us, nor is this garden created for our benefit, but it was created to fulfill God's plan and purpose. We are only observers to the unfolding of God's plan.

It is now apparent that something spectacular is about to happen. The animals seem to sense it, and so do we. Our excitement abounds as every passing moment is filled with joyous anticipation. Whatever the Creator is about to do, it will probably pale all other creative acts. All of the previous spectacular creations of heaven and the angels, the universe and its billions of stars, and the recreation of earth, has all been created for this moment. The moment is yet to be cherished, when God can walk in His garden with the greatest of all His creations: man.

IV MAN BECAME A LIVING SOUL

The Garden of God is almost dream-like in its magnificent beauty. We have only begun to explore its idyllic wonders, but any further exploration must wait, because the excitement is all around us in anticipation of something glorious about to happen. We have arrived at the grove and are stretched out under a sprawling tree to enjoy the view of the lagoon. There are a large variety of animals in the area, which thinks that we are the attraction to watch. Many butterflies flit about, wearing exotic colors and are as big as my hand, while others are smaller and transparent and yet seeming to magnify the color of the flower on which they rest. The scent from the patchwork of flowers nearby is caught up in the gentle breezes and caresses our senses. We are in sheer ecstasy, and have a peace that passes all understanding. We are truly in Paradise.

Suddenly, the animals turn in unison toward the sky to observe the glorious arrival of the WORD. The glory cloud lights up the grove more brightly than the sun, and He slowly descends to the large, weeping willow tree next to our location. The Creator gently places the limp body of the man on the green velvet-like grass beneath the welcoming arms of the weeping willow and slowly ascends above the lagoon, pausing in place to observe everything from a higher vantagepoint. The ever cautious deer inch closer, and the peacocks fold their luminous plumage out of respect for the creation of God which lies before them. The songbirds cease their charming caroling, and the Garden of God grows silent.

The silence of Eden is suddenly broken, when the WORD calls out in a loud voice; *"Adam, arise,"* and the man begins to stir from his deep sleep. Adam awakens from his deep sleep to be in the presence of his Creator. As Adam stands to his feet, his eyes are riveted upon the radiant glory before him. He immediately falls to his knees and lifts up his arms and gives praise to his Creator and his God. The animals can no longer restrain themselves, and they, too, join in the praise. It is a cacophony of raucous noise, but the Creator who is still positioned above the lagoon greatly rejoices in the noisy worship service. The WORD graciously acknowledges their praise and slowly begins to descend toward His most wonderful creation.

The Creator embraces the man, and they begin to walk together through the garden in the cool of the day. God begins to tell the man where he is and the purpose of his existence.

> *"And God said, Your name is Adam, which means 'of the ground'. I formed you from the dust of the earth and breathed the breath of life into you. You are my beloved creation: if you continue to obey my words, I will bless you forever."*

As they continue to walk together, we notice that the relationship begins to change from Creator and creation, to Almighty God and His

"Companion," or "Counterpart." It is truly a tender and intimate scene to behold, but we must draw closer to hear every word that is said. The WORD continues to recount His story to Adam about the creation of the universe and his own world.

> *"These are the generations of the heavens and of the earth when they were created, in the day that the LORD God made the earth and the heavens,*
>
> *And every plant of the field before it was in the earth, and every herb of the field before it grew; for the LORD God had not caused it to rain upon the earth, and there was not a man to till the ground.*
>
> *But there went up a mist from the earth, and watered the whole face of the ground."*
>
> *(Genesis 2:4-6)*

As they stroll through this magnificent garden, Adam is experiencing the marvelous sights of unparalleled beauty. They come upon a most extraordinary scene near the center of Eden: a graceful, cascading waterfall splashing its sparkling, jewel-like waters upon tier after tier, until the waters thunder into the waiting pool below. The scene is idyllic, and the Creator patiently answers every question that Adam poses. His fertile mind is absorbing every minute detail. Nothing escapes his notice. He queries the Creator about everything, each animal, every tree and bush, the water, and even the sun. Adam loves to be in the presence of his Creator God, and he also loves the fabulous garden that the Creator has made just for him. Everything within these placid surroundings that one can see and hear are soothing to the very soul. Adam is enthralled with it all.

V THE EDENIC COVENANT

The awesome serenity of the Garden of God is overpowering Adam. He is in sheer ecstasy, and unaware that there is a danger in Eden. The WORD turns away from the cascading waterfall to share with Adam, God's first and only law in the Garden of Eden. The LORD God turned to the man and said:

> *And the LORD God took the man, and put him into the garden of Eden to till it and keep it.*
>
> *And the LORD God commanded the man, saying, Of every tree of the garden thou mayest freely eat;*
>
> *But of the tree of the knowledge of good and evil, thou shalt not eat of it; for in the day that thou eatest thereof thou shalt surely die."*
>
> *(Genesis 2:15-17)*

The Creator points to the forbidden tree in the center of many other fruit trees, and says; *"of this tree you shall not eat the fruit thereof, lest you die."* The "tree of the knowledge of good and evil" is beautiful to look upon, for it has vibrant multi-colored leaves that shimmer in the bright sunlight. Within that leafy maze of alluring color were branches bowed low laden with delicious looking fruit. The tree is unique in its appearance, and does not resemble any of the other trees in the entire Garden of God. I must admit that it is tempting to taste of its fruit, because it is very appealing to the eye, but God has forbidden man to taste of it.

The man, Adam, agreed to obey his Creator, and not to eat of the fruit of this tree. This was the only law established by God, and a pact was made between man and God to obey that law. This agreement is known as the *"Edenic Covenant,"* it is a test of obedience. The WORD spoke to Adam, saying: *"I have to leave you for a while, go and explore more of your garden, and I shall return to you in the morning by the lagoon where you first awakened."* The Creator ascends slowly, then soars toward the late afternoon sun. Adam looks longingly at the radiant glory of his Creator God, until he can no longer see his God. Looking a bit lost, Adam gazes upon the "forbidden tree," and turns his back on it, and heads in another direction. He left behind the trees that are planted by the still waters of the pool near the waterfalls. As he walks, Adam ponders the warning that the Creator gave to him about the "tree of the knowledge of good and evil." He begins to explore the Paradise that his God has created for him. Anxious to see everything that the Creator has made, Adam set out to experience the beauty of it.

VI A SERPENT IN THE GARDEN

Adam was elated, as he went off to explore more of the serene settings in the Garden of God. This was the first time that Adam is alone without being in the presence of his Creator, but he seems to draw a sense of peace in knowing that he is obeying his God, and that his God promised to return by the morning of the next day. We could see the disappointment on Adam's face when the WORD had to leave, but the promise of His soon return lifted his heart. We both decided not to follow after Adam, but attempt to locate the WORD, for we do not want to be left behind again. Our curiosity compels us to seek after the WORD. It has been our experience, to be in the presence of the Creator God, is to be where great and wonderful events occur. We do not want to miss any opportunities to be in His presence. We rush back to the lagoon where we left our Eternity Explorer, and set out on a new adventure.

We head in the direction that the WORD had traveled. As we look back upon the radiant glow of the Garden of Eden, we are assured of God's presence upon planet Earth. The WORD had once made the entire world holy, and it glowed with the radiant glory of the Creator God, but was destroyed by Satan and his fallen angels. Now there is only a small microcosm of what once was, but also the promise of what is yet to come in the distant future. It is reassuring to our very souls that God's plan of *"Creation and Redemption"* is right on schedule and can never be stopped by any force, in heaven or on earth.

It doesn't take very long to locate the WORD, because the brilliance of His glory lit up the countryside and could not be missed. We descend slowly into the center of the brilliance, not knowing what to expect. Once we penetrate the glory radiance, we are able to see in the same manner as we did in Eden. Everything is perfectly clear, but there is a slight golden aura in this entire area. There was nothing to prepare us for the amazing scene that is before us. We rush toward the WORD, in order to hear everything that He is about to say in this most amazing setting.

The Creator has sought out the mighty warrior, fallen angels and their high commander, Satan. He has gathered together all of the leaders of the fallen race, including their infamous leader. They are all seated before their Creator in a semi-circle, with Satan seated in the midst of them, and now, the WORD begins to speak.

"You were once the mighty princes of Heaven, but because you have chosen to sin against God, you were cast out of Heaven unto the earth. The earth was one of my most beautiful creations, and you destroyed it.

I have restored the earth, and created all of the life upon it. I have created a holy sanctuary for the greatest of all my creations: man. It is my command that you shall not again destroy the earth. You shall not destroy the life upon this planet. You shall not destroy my garden, nor shall you do harm to the owner of this world, man. If my commands are not obeyed, you will surely be destroyed.

I shall now return to the Garden of Eden, and I am taking Satan with me to see the man. He shall see for himself, my glorious creation, and report back to you the truth of the matter. Heed my commands lest you be destroyed. Forget not that I am your Righteous Judge and your Holy Creator God. Obey and live."

The WORD walks over to Satan, and lays hold of the once anointed cherub, then slowly soars above the trembling gathering of the princes of darkness. Satan, in a vain attempt to save face, begins to battle and scream, but resistance against the Son of God is futile. Once they are out of sight of the fallen angels, Satan ceases his sham performance. In the distance, we can see the golden glow of God's glory shimmering in the setting sunlight. We wonder if, at the time Satan entered into the glory, would it disappear as it did in the time when the fallen angels had entered the golden aura of the original creation. Once the fallen angels were cast out of heaven and exiled to earth, the worldwide glory of God vanished upon their entrance. Now that Satan, the source of all sin, is about to enter into holy ground, will God's glory on earth, once again be lost? We will soon find out the answer, because we are about to enter the holy glory of Eden.

We land by the still waters near the cascading waterfalls, and Satan seems to be quite disturbed by the serene beauty of Eden. Perhaps he was remembering the world of beauty he had, before his rage and hatred for the WORD caused him to bring destruction upon the planet. We look about to see if the glory has disappeared, but we are elated to see that the golden aura remained. We come to the conclusion, since God is here, God's glory is here, and no power in heaven or on earth, can destroy it. This is man's planet, and as long as man remains sinless, God's glory shall remain forever.

The Son of God commands Satan to stay in this area by the cascading waterfalls, and by this grove of trees. In the center of this grove is the enticing "tree of the knowledge of good and evil."

> *"Satan, remain here for the night, and tomorrow I shall bring the man for you to observe. Take heed at what you do, for you are on holy ground."*

Suddenly, the Creator slowly begins to rise above the waterfalls and disappears over the horizon. Satan has not said anything since his display of bravado before his mighty warrior princes. Even now, he seems to be deep in thought. Perhaps, he is remembering the glorious wonders of heaven, where he was exalted as the "anointed cherub." Maybe, he is recalling the last time he saw the Son of God, who was the "Righteous Judge" at his trial. Then again, seeing all this wonder before him, he may now be regretting the destruction of the pastoral planet. He becomes keenly aware of all his surroundings, and closely observes the beautiful exotic tree in the center of the grove.

VII ADAM NAMES THE ANIMALS

The morning sunlight comes streaming over the adjacent lagoon, where we first entered Eden. The golden shafts of light give birth to a new day, and we rise up with new expectations and new wonders to behold. In a dazzling display of beauty by the perfectly named birds of paradise, and the peacocks spreading full their fabulous iridescent plumage, greet the new morning. We stand up and slowly turn about to observe our tranquil surroundings. The idyllic setting of Eden is breathtaking, and its picturesque beauty appears to take on a new freshness in the early morning light. We notice Adam beginning to stir from his sleep. He slept under the very same weeping willow tree near the lagoon, which he awoke from when he was greeted by his Creator. Adam looked for his Creator above the lagoon, but he was not there. The rhythmic noise of a woodpecker digging a hole for a nest in a nearby tree distracts our attention for a few moments, when suddenly the entire area is flooded with the now familiar golden glory of God.

Adam is elated to see his God again. He seems to be happiest when he is in the presence of his Creator and "companion." The WORD pauses above the lagoon as before, and the aura of His glory outshines the morning sun, and yet it is never difficult to look upon the face of the Son of God. The golden reflections flashing from the lagoon's rippling surface, give the appearance of the lagoon being ablaze with fire. The animal population gives a rousing welcome to their Creator. Roars, snorts, trills, brays, and trumpets broke the placid morning solitude, and the man also gives praise and worship to his Creator God.

The WORD descends to the ground, and Adam and his God begin to walk through the garden. The "companion" relationship seems to be quite an appropriate term, for they have established a bond that is closer than that of a brother. They are united in a Spiritual bond that is linked by the love of the Holy Spirit of God. The walk to their destination is a fairly short distance from the lagoon and they come upon a small clearing with a single tree in the midst. We had not seen this tree before, and we surely would have remembered it. It is one of the most unusual trees in the entire Garden of Eden. The leaves seem to be translucent, and glisten and shimmer like a gem in the morning sunlight. The natural tints and hues of the leaves seem to be engulfed in the radiant, brilliance of God's glory. There are twelve major branches to the tree, and on each branch is a different kind of delicious fruit. God said to Adam, *"This tree is called the tree of life, and you may eat of the fruit of it. It shall be for your good health and for your pleasure."* The *"tree of life"* is given to God's companion in the garden, but we remembered that the "tree of life" would also be near the river that flows from the Throne of God in the New Jerusalem as it is described in the Holy Scriptures.

"And he showed me a pure river of water of life, clear as crystal, proceeding out of the throne of God and of the Lamb.

In the midst of the street of it, and on either side of the river, was there the tree of life, which bore twelve kinds of fruits, and yielded her fruit every month; and the leaves of the tree were for the healing of the nations."

(Revelation 22:1&2)

The Creator tells Adam to remain in this location, because He is going to gather all of the animals to pass by Adam, and whatever name that he would call them, that would be their name thereafter. The Creator, once again, leaves Adam unattended, as He slowly ascends to summon the creatures of the fields, and the birds of the air, and every creeping thing. Every living creature shall pass by Adam. While waiting for the WORD to bring the animals before him, Adam decides to eat from the "tree of life," since the Creator said that he may eat of it. Adam picks a juicy looking fruit, sits down beneath the "tree of life," and begins to eat his morning meal. It is not long before we can hear and feel a rumbling of the earth. We stand to our feet, as enormous beasts come lumbering through the woods. They are the brontosaurs and the tyrannosaurs that were in the lagoon, and now slowly pass by Adam. These are the great *"behemoths"* that Job speaks of, that lived in his time, as he relates in the Bible.

"Behold now behemoth, which I made with thee; he eateth grass as an ox."

(Job 40:15)

The great beasts slowly move off, grazing in the lush fields, as they go. The carnivores and the herbivores both eat grass at this time, because there is no sin and no death in the world. The large animals were to pass by first, so that they would not trample on the smaller ones. God is a God of order and logic. He also cares even for the least of His creations. The creatures pass by Adam, two by two, the land animals, then creeping creatures, and finally the birds of the air come and land before him. They sing, warble, and trill to give honor to Adam, as he names each and every creature that passes before him this day. We are astonished at the perfect mind and wisdom of Adam. His mind is the perfection of the Creator. The more time that Adam spends with his God, the more his mind becomes like his God's. The day is well spent, as Adam names the last of the creatures. Thousands of species pass by Adam, and he is pleased to see each one, and rejoices in the wonderful creations of his Creator God. Adam rested under the "tree of life," and marvels at the many animals that remain nearby. A pure white lamb is lying on the velvet-like grass, near a patch of exotic flowers. A male lion slowly walks over to the lamb and lays down beside it. Adam, the lion, and the lamb are all created on the last day of Creation. They all represent the completion of God's plan of *"Creation."* The lamb is symbolic of the completion of God's plan of *"Redemption."* The lion is a prophetic symbol of the resurrected and glorified Jesus, the Son of God, who is the *"Lion of Judah."* He will rule the world from the Throne of David, for one thousand years, as *"King of Kings"* and the *"Lord of Lords."* What a wonderful prophetic scene that we are privileged to witness in this tranquil setting of Eden. We are rejoicing to be so richly blessed by the presence of such beauty and serenity and hope that it can last forever, but will it?

Chapter 8

And God said, Do not eat of it, lest you die

It was a very exciting day for everyone, so we tasted some of the exotic fruit of God's Garden, and we quickly discovered that we have never tasted anything so delicious. We lay down to rest and enjoy the serenity, spending the night near the "tree of life." We observe Adam, who appears to be troubled about something. It is a peaceful night, and sleep is a welcomed guest. We marvel at the millions of stars that are the roof over Eden, as we drift off to sleep. We are awakened by the cooing of morning doves, and the choir of songbirds who sing their greeting to the morning sunlight that is beginning to peek over the distant hills. Another gorgeous sunrise is here for us to enjoy. Adam awakes shortly after us, and seems to also be blessed by the spectacular sunrise. After enjoying a delicious meal from the nearby trees, Adam sits down and soon becomes deep in thought. His face reveals a sadness, but his countenance immediately changes to delight, when he notices the brilliant glow in the skies, coming toward us. The WORD greets all of us, as Adam and his God walk together in the cool of the morning. The Creator quickly becomes aware of Adam's sullen countenance and his saddened heart. Adam says not a word, but his loving God knows the heart of man.

> *"And the LORD God said, It is not good that the man should be alone; I will make him an help fit for him."*
>
> *(Genesis 2:18)*

I THE CREATION OF EVE

When Adam was naming the animals, he was quick to notice that the Creator had made each creature male and female. Each of the creations that passed by Adam had a "counterpart," but the Creator did not have a "counterpart," nor did Adam have a "counterpart." Adam's heart reveals his innermost desires to the LORD God. The Creator is quick to respond to the needs of His greatest creation.

> *"And the LORD God caused a deep sleep to fall upon Adam, and he slept: and he took one of his ribs, and closed up the flesh instead thereof;*
>
> *And the rib which the LORD God had taken from the man, made he a woman, and brought her unto the man."*
>
> *(Genesis 2:21&22)*

The Creator causes a deep sleep to come upon Adam. He then opens an incision in Adam's side, whereby He removes one of his ribs. The incision is then closed, and Adam remains in a deep sleep. The Creator moves away from the still body of Adam, and scoops up handfuls of the rich earth, and begins to mold and form the earth about the rib of Adam. The glory of the Creator, once again, is difficult to observe through the radiance. This creation is an act of love for Adam, so that Adam will never again be lonely. When the Creator has finished His creation, He breathes the breath of life into the body lying upon the ground. The creation stirs and sits upright, as she gazes upon her Creator. She raises her hands in praise and worship. With a gentle voice, the Creator speaks to her, saying: *"You shall be called Woman, because you were taken out of Man."* The woman is beautiful beyond description. Every line and curve of her body is perfect and without flaw. Her countenance is radiant with love for her Creator God. The glory of God still remains upon her. The Creator brings her to the man, and as they gaze upon one another, elation and love fills their hearts and they begin to walk through the beautiful garden together. Disappearing out of sight, we notice a rustling in the bushes, like someone lurking behind the trees watching everything that occurred. The observer did not want his presence to be made known. It is a great mystery, but there does not seem to be anyone there now.

We have momentarily been abandoned in this grove, so let's discuss what has occurred here. We could state that the first Adam gave birth to his "Bride" from the gash in his side, in Eden. While in the future, the second Adam (Jesus) will give birth to His "Bride," the Redeemed, from the gash in His side, at Golgotha. The "Bride" was given the first Adam, as a gift from God. (Genesis 2:22) In the distant future, the "Bride" will be given to the second Adam (Jesus), as a gift from God (John 3:16). The birth of the 'Bride" in the Garden of Eden was a physical birth, while the birth of the "Bride" at Golgotha, will be a Spiritual birth. The marriage of the creations of God, in the Garden of Eden will be the culmination of God's plan of *"Creation,"* while the marriage of the Redeemed to the Son of God, at the Rapture, will be the culmination of God's plan of *"Redemption."*

ADAM AND EVE MARRIED

The morning brings a fresh new day, filled with great joy and anticipation of exciting events to come. The creatures of Eden are in an excited state, and the birds are welcoming the new day with joyous song. It is another glorious morning in Eden, with its stunning beauty everywhere one goes. Adam had lovingly built a small lean-to for the woman, under a nearby tree, to protect her from the morning dew. Adam slept close to our camp, but our presence does not appear to bother them. The sudden raucous call of the animals awakened all of us at once. The noisy excitement of these creatures can mean only one thing. The WORD is coming. His creations seem to sense the coming of the Creator every time He appears. They certainly know how to awaken a person from a comfortable sleep.

The glory of the Creator can be seen coming over the distant hilltops. It is like having two suns splashing over the horizon from different directions. We never tire of beholding this magnificent sight. All of Eden greets their Creator with loud praises. The WORD graciously acknowledges their praise as He lands and greets the man and the woman. They begin to stroll together in the cool of the morning, toward an area that God had prepared. We enter a glen, which has grass like velvet, and a walkway lined with arrays of exotic flowers. At the end of the pathway are two trees, with their tops bowed toward each other, forming an arch. The Creator God places the man and the woman, under the spreading arms of the trees, and He pronounces them to be man and wife, and blesses them under the sanctity of holy marriage.

> *"And Adam said, This is now bone of my bones, and flesh of my flesh; she shall be called Woman, because she was taken out of Man.*
>
> *Therefore shall a man leave his father and mother, and shall cleave unto his wife; and they shall be one flesh.*
>
> *And they were both naked, the man and his wife, and they were not ashamed."*
>
> *(Genesis 2:23-25)*

The many creatures that had gathered to witness that great event had no idea of what had just taken place. There was however, a figure lurking in the shadows of the trees that seemed to take great interest in these proceedings. We just realized that there were countless numbers of beings that were observing this wonderful event. The "Angels of God," in heaven, have been watching the entire plan of God unfolding, from the beginning. Their understanding of the creation of Man as God's "companion," probably was not a difficult concept to grasp. However, the concept of a "female counterpart," is not known in all of heaven, or in the universe. The Creator God has no "counterpart." The angels of heaven have no "female counterpart." The "Age of Creation" concludes with the creation of the 'female counterpart" to man, and the "Age of Innocence" begins.

Unknown to the curious angels of God, in heaven, the creation of the Man and the Woman, was the beginning of a great new adventure. God created the man as a "companion" for Himself. God created the woman as a "companion" for the man. The mystery of the ages is that the man and the woman are the progenitors of the "Bride of God," or the "counterpart" of God, in the distant future. The Bride, or counterpart of the Creator God, will be the "Redeemed" who have believed and trusted in God, throughout the ages. At an appointed time in history, God will claim His beloved counterpart to be with Him forevermore throughout eternity.

Perhaps, the angels in heaven have understood this concept, because of the speech that God gave to them in heaven, explaining His plans of *"Creation"* and *"Redemption."* We do believe, however, the mysterious being that has been skulking in the shadows of the glen, has an understanding of all of the beautiful events that have occurred. The shadowy figure seems to take great interest in everything that occurs, with the Creator God and His creations.

The holy marriage ceremony is now completed, and the WORD begins to walk with the newly married couple, through the glen toward the highest mountain in all of Eden. To see the ecstasy in the eyes of this loving couple is thrilling. The institution of marriage was created by God and was blessed by God. It is truly unfortunate that so many people of our time think that marriage is unnecessary and deems the rite meaningless. Some are married by government officials to satisfy the law, but totally miss out on the holiness of the ceremony and the blessings of God. There is a reason for the traditional ceremony to be called "Holy Matrimony." God's holiness, God's blessings, and God's law, are all involved in the union of two souls and two spirits, to be joined as one, in holy matrimony.

II TITLE DEED TO EARTH GIVEN TO MAN

We have finally completed the long climb to the summit of the highest hill in Eden. The golden aura of God's glory is more apparent at the higher elevation, which makes the breathtaking view even more spectacular. In a moment of silence, we all behold this extraordinary panorama. We can see a great expanse of forests with all manner of trees. There are hills, and rivers, with broad cascading waterfalls, roaring in the distance. In every glen, there are multitudes of exotic flowers that exude a variety of bouquets of delicate scents. The Creator waves His arm in a circular manner, to indicate that the entire expanse of His glorious creation belongs to the Man and the Woman.

> *"And God blessed them, and God said unto them, Be fruitful, and multiply, and fill the earth, and subdue it; and have dominion over the fish of the sea, and over the fowl of the air, and over every living thing that moveth upon the earth.*
>
> *And God said, Behold, I have given you every herb bearing seed, which is upon the face of the earth, and every tree, in which is the fruit of a tree, yielding seed; to you it shall be for food.*

And to every beast of the earth, and to every fowl of the air, and to every thing that creepeth upon the earth, wherein there is life, I have given every green herb for food: and it was so."

Genesis 1:28-30)

The WORD has just given to man all that He has created. Man is to have "dominion," or total control, over every plant, every creature of the seas, and every creature of the earth. The Creator God has given an awesome responsibility, to literally rule the world.

"All things were made by him; and without him was not anything made that was made."

(John 1:3)

Everything that the WORD had created upon the earth, including the earth itself, now belongs to man. The "Son of God" has given the "Creation of God," as an inheritance, to the "Child of God." Man has inherited the earth and all that is in it, and will be the "Heir of God," only as long as he continues to be the sinless, eternal image of God.

III PARADISE IN EDEN

In this *"Age of Innocence,"* the Man and the Woman are in a perfect relationship with their Creator God. They never want to be away from His presence, and are always seeking His council, and His blessing. They have given praise and worship to their Creator God at every opportunity. Their heart-felt desire to be near to their God, and to be obedient to His

Word, is always rewarded by blessings. Their lives are abundantly rich and full, as man and his wife walk in the center of God's will. They share the experience of blissful ecstasy, in their love for each other, and a love for their Creator God that passes all understanding. The loving Holy Spirit of God, is the common link, between the man and the woman and their LORD God. The Creator breathed life into the man and the woman, by imparting the Holy Spirit of God into their lifeless bodies. The love of God, and the Spirit of God dwells richly within them. Eden truly is a fantastic Paradise. The joy that is evident here is beyond words or feelings.

As we take one long last look, at this tranquil panorama, the WORD tells the newlyweds to *"Explore your home of Eden, and remember all that I have told you. I shall return to you each day, and we shall walk together, in the garden, in the cool of the day."* With a final loving look to the man and the woman, the WORD slowly ascends, and pauses to look back one more time, upon His beautiful creations. The radiant glory, of the Creator, slowly disappears over the horizon. It is like a spectacular sunset, but the sun is still high above us. Even though it appears to be difficult for the Creator to leave His "companions," He is giving them time alone to nourish and develop their relationship with one another.

The elated couple embraces one another, and they both stare longingly, as the last traces of the Creator's glory disappear from sight. The couple turns to leave the hilltop, but takes one more look back, in hopes of getting one more glance at their Creator God. Adam lovingly gazing upon his beautiful wife decided to give an appropriate name to her. The Creator said they were to rule the world and have dominion over all the creation.

"And Adam called his wife's name Eve, because she was the mother of all living."

(Genesis 3:20)

We begin the descent; down a tree lined pathway that exudes exotic scents from the flowers beneath their sturdy limbs. Birds of every description burst out in a song of blessing to Adam and his bride, Eve. The couple stops, and Eve holds out her hand, as two small songbirds land on it, and sing a lilting song to the lovely woman. She kindly thanks them, as they fly off to a nearby tree. Loving creatures greet them everywhere they go. Adam takes Eve to see all the wonders of Eden that they can explore in one day. They visit springs and streams, waterfalls, and ponds, and find artesian fountains majestically shooting their waters high into the air. It is a joyous time they share this day, as the approach the grove of trees that is by the cascading waterfalls and the still waters.

Adam and Eve rest under the trees by the still waters. Then after a short rest, they wade into the still waters. They laugh and giggle, as they splash each other in the crystal clear waters. Many of the creatures nearby gaze upon the couple that are celebrating life together. Some of the animals enjoy the frolicking in the water so much; they too jump in the water, and join the fun. Deep in the shadows there is another creature that is observing the entire joyous scene, but only cowers back deeper into the darkest abyss of the grove.

Adam and his wife wade out of the still waters and again rest upon the velvety grass, under the sprawling branches of a large tree near the water's edge. They are now hungry and look about for fruit trees, to eat of their fruit. Still laughing about the fun time in the water, they gather many varieties of delicious fruit to eat. Adam remembered the Creator's command and took Eve to the center of the grove, to the beautiful tree that God had showed to him. Adam told her about God's command and warning about this tree.

> *"And the LORD God commanded the man, saying, Of every tree of the garden thou mayest freely eat;*
>
> *But of the tree of the knowledge of good and evil, thou shalt not eat of it; for in the day that thou eatest thereof thou shalt surely die."*
>
> *(Genesis 2:16&17)*

Eve gazes at the *"tree of the knowledge of good and evil,"* for it was enticingly beautiful to behold. Its multi-colored leaves shimmer in the gentle breezes, and the leaves seem to change color, as they vibrate in the soft breeze. How inviting the fruit of this tree appears to her, as Eve's curiosity draws her closer to the tree. Suddenly, there is a movement in depths of the darkness of the tree, and Adam grasps her arm. Adam reminds Eve of the Creator God's commandment about this particular tree. They turn from the "forbidden tree," and head for the lagoon, laughing and rejoicing in their love for their peaceful home in Eden, and for their love for each other.

IV SIN ENTERS PARADISE

Time passes slowly in the Garden of Eden. The happy couple shares a pure love for each other and for their God. They explore each place in

Eden, with joy and enthusiasm. Together, they play with the animals, as though they are pets. The wonders in Eden are endless. The WORD visits the man and the woman, each day, and walks with them in the cool of the day. Their bonds of "companionship" grow each day, as they share with one another the deep things of God's mysteries.

One beautiful morning, Eve rose up early, and is careful not to awaken her husband. She decides to gather some fruit, as a surprise, for her beloved husband's morning meal. Shafts of morning sunlight are piercing through the trees, and the morning dew glistens from every flower and bush. Branches of trees wave enticingly in the warm gentle breezes. Eve searching out new and exotic fruit to surprise her husband, sees the familiar cascading waterfalls by the still waters, and pauses to hear its roar, as the waters come crashing down. Tumbling into rocks below, the waters become lifeless and still. This spectacular sight is beautiful, particularly at the dawning of a new day. Eve recalls another beautiful sight that is close to the waterfalls, the unusual tree in the center of the grove, the "tree of the knowledge of good and evil."

She approaches the tree cautiously, and is enraptured by its beauty in the bright morning sun. The multi-colored leaves shimmer enticingly in the gentle breeze. The branches wave a beckoning call to come closer. Eve marvels at the alluring fruit that is ripe for the picking, and appears to be delicious to the taste. The colors and shapes of the fruit are unlike any other fruit in all of Eden. Eve draws closer to the "forbidden tree," and is startled by something slowly moving, deep in the shadows of its branches. Out of the darkness came a soft alluring voice that speaks to her.

"Now the serpent was more subtle than any beast of the field which the LORD God had made. And he said unto the woman, Yea, hath God said, Ye shall not eat of every tree of the garden?

And the woman said unto the serpent, We may eat of the fruit of the trees of the garden;

But of the fruit of the tree which is in the midst of the garden, God hath said, Ye shall not eat of it, neither shall ye touch it, lest ye die."

(Genesis 3:1-3)

Eve is captivated at the opportunity to talk to someone new in Eden. Coming from the depths of the darkness, the soft voice is now coming closer to the woman. She draws even closer to the tree dazzling in the sunlight. This new being of darkness, has now moved toward the woman, and is now partially in the light, as he speaks once again to her, in a soft beguiling voice.

"And the serpent said unto the woman, Ye shall not surely die;

For God doth know that in the day ye eat thereof, then your eyes shall be opened, and ye shall be as God, knowing good and evil.

(Genesis 3:4&5)

The serpent, a writhing reptile in the garden, is not the present form of Satan, but God calls him a serpent, because the serpent may have been the most graceful and subtle of all the creatures in Eden. The term "serpent" has remained in the Scriptures to depict the subtlety of his deceptions, for the destruction of mankind.

> *"And the great dragon was cast out, that old serpent, called the Devil and Satan, who deceiveth the whole world; he was cast out into the earth, and his angels were cast out with him."*
>
> *(Revelation 12:9)*

The serpent, Satan, began his conversation with the woman, by speaking the truth of God's command, but quickly interjected a bold-faced lie stating "Ye shall not surely die." Since Satan is a liar, and the father of lies, it is now one of his chief weapons of destruction. Satan now comes out from the darkness of lies against the Creator God, and assumes a role as an "Angel of Light." He not only presents himself as an "angel of light," but with cunning, he presents himself to be an "Angel of THE Light."

> *"And no marvel; for Satan himself is transformed into an angel of light."*
>
> *(2 Corinthians 11:14)*

Satan is quite blatant here in offering the woman information that her God has not told her. Satan has given her a cunning half-truth. To "know good and evil," one would not be like God, but one would be like Satan himself. He did not tell her, whosoever sins against their Holy God, would be condemned, eventually, to die not only physically, but spiritually, separated from God forever and ever, in the "Lake of Fire." Satan knew this as fact, because he heard God's speech on His plan of *"Creation and Redemption,"* in heaven. Satan's cunning plan is to deceive Eve, with truths, half-truths, and lies, twisting God's Word, to serve his own dark purposes.

THE FALL OF MAN

Adam awakens to the cheerful singing of songbirds overhead. He becomes concerned that he cannot find his beloved wife as he calls out to her, but to no avail. Adam then begins to search for Eve in some of their favorite places, where they shared so many joyous times together. Soon, he approaches the beautiful cascading waterfalls by the still waters. He pauses for a moment, to reflect upon the glorious times that he had spent there with his lovely bride. Turning, he sees Eve close to the tree in the center of the grove, and rushes to her side.

Adam embraces his wife, relieved to see that no harm has come to her. He immediately notices the being in the shadows of the forbidden tree. This lurking figure is partially in the darkness, and partially in the bright light. Eve had gathered much fruit for their meal, and had placed the fruit nearby at her feet. She is holding in her hands, the alluring fruit from the forbidden tree. At the assurance of the being in the shadows, that the fruit would *"make one wise,"* she had eaten of the fruit. Eve had eaten of the fruit of the "tree of the knowledge of good and evil." Her eyes were opened to that knowledge of evil, for she already knew good, because she was created perfect, and was a "companion" to the Living God. The "Prince of Darkness," offers the woman the light of illumination, of not only knowing the mysteries of God, but also knowing the dark mysteries of the damned.

Adam is stunned at the sight of his mate eating of the fruit that the Creator God had condemned. Her eyes were opened to the knowledge of good and evil, and her first opportunity to do evil was before her. She offers the enticing fruit to her husband, for it is very pleasant to look upon. Eve gives the forbidden fruit to her husband, for, she says it will *"make you wise,"* and he eats.

> *"And when the woman saw that the tree was good for food, and that it was pleasant to the eyes, and a tree to be desired to make one wise, she took of the fruit thereof, and did eat, and gave also unto her husband with her; and he did eat."*
>
> *(Genesis 3:6)*

The Prince of Darkness shrieks in delight and steps back from the bright light of the day, disappearing into an abyss of darkness. With the sound of the serpent's sinister laughter still ringing in their ears, the loving couple turns to look at one another. Their eyes now being opened, they knew that they are naked, and they bow their heads in shame. The *"Age of Innocence"* is now over.

> *And the eyes of them both were opened, and they knew that they were naked; and they sewed fig leaves together, and made themselves aprons."*
>
> *(Genesis 3:7)*

At this point, was there anything different that they could have done? Let's investigate some of the options.

- Adam seeing that his wife had eaten of the "forbidden fruit," could have said, "Eve, you have disobeyed the Creator, let us go together to Him and seek His mercy, and ask for forgiveness."
- Adam could have refused the "forbidden fruit," and asked the Creator to censure him, because he was the head of his household, and is responsible for his family, and the covering for his wife.
- Adam could intercede, with the Creator, to save his wife, and offer himself as the object of God's punishment.

As we can see, Adam had other options to pursue, but chose to disobey the commandment of the Creator God anyway. The Bible is silent, and gives no reason for Adam's disobedience to God's only law in Eden. Perhaps, he loved his wife so much, he willingly chose to share with his beloved whatever punishment their God meted out. What do you think would have happened, if Adam refused to disobey his Creator, and had not eaten of the forbidden fruit?

V THE KNOWLEDGE OF GOOD AND EVIL

The life of ecstasy in the idyllic setting of Eden, is now doomed. The perfect couple for the first time of their lives feels shame and fear. They hurry over to a nearby fig tree to gather its leaves and sew them together with the grass beneath the tree. Panicking, they dread seeing their LORD God. With their "eyes now being opened" to good and evil, they know that evil cannot be in the presence of a Holy God. Their fate is now uncertain, and fear confronting the Creator, so they hide themselves.

Their "eyes have been opened to good and evil," and now fear confronting a Holy Righteous God, for they have done evil.

Suddenly, a golden brilliance fills the heavens, as the WORD slowly soars above the cascading waterfalls. Dancing waters at the top of the falls, shimmer with life, like heavenly, liquid gold, and race furiously toward the rocks below, screaming and roaring into the lifeless still lake. When the presence of the LORD God comes near the lifeless still waters, they become a golden boiling cauldron, writhing with anguish. The troubled waters seethe in torment, the flaming colors making the once tranquil, still waters, look like a "lake of fire."

The Creator seeks His greatest of creations, and knowing where they are calls out to Adam and his wife, but gets no response.

> *"And they heard the voice of the LORD God walking in the garden in the cool of the day: and Adam and his wife hid themselves from the presence of the LORD God among the trees of the garden.*
>
> *And the LORD God called unto Adam, and said unto him, Where art thou?*
>
> *And he said, I heard thy voice in the garden, and I was afraid, because I was naked; and I hid myself*
>
> *And he said, Who told thee that thou wast naked? Hast thou eaten of the tree, whereof I commanded thee that shouldest not eat?"*
>
> *(Genesis 3:8-11)*

Was Adam trying to protect his wife here, or was he only looking out for himself? Notice that Adam's response is always in the first person; *"I heard, I was afraid, I was naked, I hid myself."* His reply gave the appearance that Adam was protecting his wife, by not even mentioning her. Perhaps it was a vain attempt to shield blame from his wife. But in his next reply, Adam seemed to have second thoughts, and tried to save himself, by blaming the woman.

> *"And the man said, The woman whom thou gavest to be with me, she gave me of the tree, and I did eat.*
>
> *And the LORD God said unto the woman, What is this that thou hast done? And the woman said, The serpent beguiled me, and I did eat."*
>
> *(Genesis 3:12-14)*

The woman replied truthfully, even though she is now in a sinful state. The serpent did beguile her. Adam, on the other hand, even tried to blame God for giving him a faulty woman; consequently, God shifts His questioning to the woman, who gives Him the truth. Even though the woman was tricked into sinning by Satan, the man willingly ate the "forbidden fruit."

When Adam sinned, he forfeited the right to the title deed of the Earth, to the serpent, Satan. There will never be a man worthy to reclaim the Earth, until the glorified, resurrected Son of God will open the title deed, or the scroll, for He alone will be worthy. The Lion of Judah, Jesus, the Lamb of God, will be the only one to be found worthy, by the Redeemed, to break the seven seals of judgment upon the sinful earth (Revelation 5:5-10). The opening of the seals of the scroll will begin the seven-year Tribulation period of judgment. At the end of the seven years, the LORD God will return to the Earth with the Redeemed, and the angels of God, to reclaim the title deed to the Earth from Satan, at the Battle of Armageddon. Satan will then be cast into the "bottomless pit," and the title deed to the Earth will now belong to the Redeemed, as an inheritance to last forever and ever.

VI THE ADAMIC COVENANT

The Creator looked longingly at His beloved creations, and with sadness in His heart, knew what had to be done. Even though the LORD God knew what would happen, even before the heavens were created, His heart broke at the thought that His beautiful creations, would no longer be His "companions." There could be no more strolls through the pastoral garden, sharing their love for one another. The holy "companions" of God, were no longer holy. The perfection of their creation had now been tainted by sin and the aura of God's glory no longer radiates from them. Their sin has separated them from their LORD God, and the trusting relationship they once had, is now replaced with fear of their Creator. Sin always separates man from God, but in His plan of Redemption, God has made provision to cancel the penalty of sin through the shed blood of Jesus, the Son of God. Adam and Eve will not be redeemed until the Son of God, in the distant future, pays the penalty for sin, on a cruel, bloody cross, on a hill called Golgotha.

The Son of God seeks out that old serpent, called Satan and lays hold of him as he cowers in the dark shadows of the tree of the knowledge of good and evil. Bringing him before the man and the woman, he issues forth an unusual, prophetic command.

> *"And the LORD God said unto the serpent, Because thou hast done this, thou art cursed above all cattle, and every beast of the field; upon thy belly shalt thou go, and dust shalt thou eat all of the days of thy life.*
>
> *And I will put enmity between thee and thy seed and her seed; he shall bruise thy head, and thou shalt bruise his heel."*
>
> *(Genesis 3:14&15)*

This is the first prophecy in God's Word. The LORD God curses Satan, to be forever considered, as being lower than the lowest created creature on earth. He once was the greatest of all created beings, Day Star, the "Anointed Cherub," and the "Covering of the Throne of God." As Satan, he has destroyed the relationship between the Almighty God and His greatest creation, man. Satan's fall from grace is enormous. The archangel, Day Star, was created to be God's glorious Light in heaven, but as Satan, he will ultimately spend eternity in the utter darkness of the Lake of Fire.

The second part of this marvelous prophecy, we discussed previously, about the future conflict between *"thy seed,"* the fallen angels that Satan spawned, and the *"seed of the woman,"* which refers to the virgin birth of the Messiah, the incarnate Son of God. The battle between Satan and the Son of God will take place on an old rugged cross on Golgotha. The death of Jesus, and His resurrection from the dead shall defeat Satan. Satan's days will surely be numbered because we notice the phrase, *"all the days of thy life."* Satan's days are limited, before he is cast into the "Lake of Fire." This covenant, between God, Adam, and Satan is not for Satan's benefit, but instead, for the encouragement of Adam and Eve. Their "seed" will bring forth the "Anointed One of God," to be the "Deliverer" of all mankind. This promise of God is known as the "Adamic Covenant"

Satan is cursed, and the promise of "Redemption" has been proclaimed, but the WORD now turns to the punishment of the man and the woman. The curse of man and the earth is part of the "Adamic Covenant," and the inheritance of the act of sin is about to fall upon the man and the woman for all mankind, until God's merciful plan of "Redemption" is complete. The LORD God continues the Covenant.

> *"Unto the woman he said, I will greatly multiply thy sorrow and thy conception; in sorrow thou shalt bring forth children; and thy desire shall be to thy husband, and he shall rule over thee.*
>
> *And unto Adam he said, Because thou hast hearkened unto the voice of thy wife, and hast eaten of the tree, of which I commanded thee, saying, Thou shalt not eat of it: cursed is the ground for thy sake; in sorrow shalt thou eat of it all the days of thy life;*
>
> *Thorns also and thistles shall it bring forth to thee; and thou shalt eat the herb of the field;*
>
> *In the sweat of thy face shalt thou eat bread, till you return unto the ground; for out of the ground thou wast taken: for dust thou art, and unto dust shalt thou return."*
>
> *Genesis 3:16-19)*

The curse of God is upon the woman, the man, and upon the earth. Let's review the facts of this curse upon mankind, and the curse upon the planet Earth.

- First, God cursed the woman, for she was the first to sin. The woman is to endure agony and suffering in birthing children. God also reaffirmed that the man was the head over the woman.
- Second, God cursed the man, because he listened to his wife and sinned, and because of his sin, the very ground that Adam was created from, is cursed. Man will now have to work hard to produce food that once came freely in the Garden of Eden. Man will now die and return to the earth, and once again, become the dust of the earth.
- Third, God's curse included the earth itself. Since Adam came from the earth, it too is cursed, and shall bring forth weeds, thorns, and thistles. The planet will never again be like the Garden of Eden, or like the restored planet outside of Eden, until God's plan of *"Redemption"* is completed.

The Scriptures tell us that the whole creation, including planet Earth, groans in pain waiting for the redemption of the Sons of God. After the Sons of God, or the Redeemed, are taken to heaven at the Rapture, and the sinful God rejecting people of the earth, are punished during the Tribulation, then the earth, too, will be redeemed. The earth will be changed to be just like the recreated earth, at the time of Eden. God tells us in His Word.

> *"For the earnest expectation of the creation waiteth for the manifestation of the sons of God.*

Because the creation itself also shall be delivered from the bondage of corruption into the glorious liberty of the children of God.

For we know that the whole creation groaneth and travaileth in pain together until now."

(Romans 8:19,21&22)

One day, in the distant future, man will be redeemed from the curse, and the earth will be redeemed to the beauty that was at the time of the restoration. This is little comfort to Adam and his wife, as the realization of the curse now begins to impact their thoughts, and their very souls. Separation from their Creator God, is a horrifying thought and the final consequence of their sin, death, is unknown to them. God told them: *"The wages of sin is death..."* (Romans 6:23a).

VII A BLOOD SACRIFICE IN EDEN

Adam and Eve stand before their Creator and Satan, realizing the full impact of sin upon their lives. The sudden realization that they are naked, is not their sin, but choosing to believe Satan, and disobeying the LORD God, was the true sin. God's Word is always true, and never to be taken lightly. The consequences of disobedience are eternal. Knowing that they were naked was only a symptom of their sin, and their first response to their nakedness, was to cover it up by their own works. The covering of sin by works is an abomination to God, and is never accepted. Because we are dead in our sins, God requires a blood covering for sin, to give us life. God's Word tells us that *"Life is in the blood."* This is why, under the Dispensation of the Law, Israel will be required to sacrifice animals, so that their blood will "cover" their sins temporarily. In the Dispensation of Grace, the Son of God will be the blood sacrifice, for all time. His blood will not just "cover" sin, but will "cancel" sin, forever.

Adam heard all that the LORD God had said, particularly that his wife would have to bear children in pain and sorrow. He lovingly turns to his beautiful wife, and softly caressed her face, knowing that some day his wife will endure terrible agony. Eve weeps seeing the anguish in her beloved's eyes. Her body trembles as she reflects on her terrible decision to believe the lies of Satan, and they embrace. Satan seems pleased with himself, because he thinks that he has finally destroyed the Creator's plan to have a "counterpart," that would forever be God's companion, because Adam and Eve are sentenced to die in a sinful state, and can never be redeemed, just like himself.

The Creator looks at the couple's feeble attempt to cover their sins, so the Son of God ascends above the small group, as they are left glaring at one another. Satan laughs out loud, and takes pride in his accomplishment. The beautiful couple is ashamed that they have sinned against their holy God, and there is no way to ever restore the relationship with God. Their thoughts momentarily drift back to their walks and conversations with LORD God, in the cool of the day. Suddenly, the heavens are filled with the golden glory of the Creator returning to them. As He descends, they all try to back away from the glory, fearing that something dreadful might happen to them. When He stands before them He gives them coats of skins for their covering. Adam looks at the bloody skins, and recognizes the pure white wool, like the "lamb" that lay next to the lion in the grove. Because of their sin, innocent lambs had to be slain. The pure, white wool of the skins was horribly stained with the sinless blood of the lambs. Adam weeps. The Holy Righteous God, shed the blood of innocent lambs to "cover" their sin, but it is in the mind of God the Father, to one day shed the blood of His only begotten Son to "cancel" sin. The death of the innocent "Lamb of God" will be the price that the Son of God will pay for all sin, forever.

> *"For Adam also and his wife did the LORD God make coats of skins, and clothed them."*
>
> *(Genesis 3:21)*

Death has entered the tranquil Garden of Eden. The earth has never known death before, not in the outside world, nor in the Garden of God. The sacred ground of Eden is no longer holy. The golden radiance of God's glory in the Garden of Eden has disappeared. Sin and death have entered into holy ground, and now instead of being blessed, Eden is now under the same curse of God, as the rest of the earth, because of the sin of Adam and Eve. The idyllic beauty of the garden has lost a lot of its mystique and serenity, and is beginning to look much like the outside world. We can hear animals off in the distance by the placid lagoon, growling and snarling at each other. The animals that were once so trusting and gentle are now hiding from one another in fear. We can see in the distance, an animal that only yesterday, was eating grass is now eating another animal that it has just killed. The Garden of Eden was God's wonderful gift to man, and the title deed to the earth. Both are now truly in Paradise lost.

VIII ADAM AND EVE CAST OUT OF EDEN

The Creator God sadly looks at his beautiful creations, then turns to Satan and speaks softly to him, or perhaps the WORD is speaking to the other members of the Triune Being. It is not easy to determine to whom the statement is directed, but it is an emotional and difficult command for the LORD God to give.

> *"And the LORD God said, Behold, the man is become as one of us, to know good and evil; and now, lest he put forth his hand, and take also of the tree of life, and eat, and live forever;*
>
> *Therefore the LORD God sent him forth from the garden of Eden, to till the ground from where he was taken."*
>
> *(Genesis 3:22&23)*

The LORD God, heart-broken, turns toward his two beautiful creations, and commands them to leave the Garden of Eden. They can not remain in the garden, because if they remain and eat from the "tree of life," they would be able to live forever in a sinful state. If Adam and Eve lived forever in a sinful state, then they would also never be redeemed. If they lived forever, and could not die, then God's command that *"if a soul sinneth, it shall surely die,"* would not be true. We must conclude that the expulsion of Adam and Eve was an act of love by the LORD God so that his beloved "companions" would one day be redeemed on that glorious day. That glorious day when the crucified Jesus will go down into Hell, and take *"captivity captive."* The souls and spirits of those who will die trusting in God, from Adam and Eve to the time of the crucifixion, will be taken to heaven. They shall all be redeemed.

Adam and Eve behold their LORD God for the last time, tears streaming down their cheeks, as they turn to leave their home. It grieves them terribly to know that they caused so much sorrow in the heart of their Creator. They had only walked a short distance, when they turned to look upon the face of God again, but He was not there. The roar of the cascading waterfalls could not drown out the soul, piercing, shrieking laughter of Satan. He is standing by the lifeless, still waters, silhouetted in front of the beautiful, alluring, cascading waterfalls, the embodiment of all evil. His triumphant laughter penetrates their very souls, and they became filled with pain and guilt.

Adam and Eve turn to run from their adversary, when Adam cries out in agony after running into a bush that had now grown long, jagged spear-like thorns. Thorns plunged deep into Adam's side. Excruciating pain surging through his body, he cries out in agony, as he quickly pulls a thorn from his side. He pressed his hands over the deep wounds to stop the bleeding and to ease the pain. Eve was quite concerned for her husband, for they have never before experienced pain, nor have they ever seen their own blood. The pressure that Adam applied to the deep wounds in his side could not stop the bleeding that flowed like water. Adam looks at his blood stained hands, and gazes upon the crimson blood for a long time. Adam wipes the blood on his hands on the skins of the clothing that God has made for him. Adam's blood mingles with the blood of the lamb that the LORD God has sacrificed for his sin, as Adam and Eve reluctantly move toward the boundary of the Garden of Eden. With the gloating, hideous laughter of Satan still ringing in their ears, the couple now reaches the end of Eden.

In obedience to the command of God, they cross over the river that is the boundary, at a shallow ford. They looked back at their home, and are startled at sight of magnificent beings guarding the Garden of Eden. The golden glory of God radiates from them, and they wield flaming swords to protect the garden.

> *"So he drove out the man; and he placed at the east of the garden of Eden cherubim, and a flaming sword which turned every way, to guard the way to the tree of life."*
>
> *(Genesis 3:24)*

The spectacular scene before Adam and Eve finally makes sense to them. These beings must be angels of God, and they were brought from heaven by the Creator, not to protect their home, but to keep them from ever returning. Finding some large boulders to sit on, under a huge sprawling tree, they watch the evening sun slowly setting over Eden. It was an extraordinary sunset for them. They had never seen a sunset without seeing it through the golden aura of God's glory. They have never seen a sunset from outside of the holy ground of Eden. Nevertheless, it is a spectacular sight, with the shadows of the hills and trees of Eden, now reaching out to the river. Contrasting those dark shadows, are the radiant glories of the cherubim with their flaming swords. The river appears to be a great gulf that separates them from where they were in the world, and their home in Paradise. The great gulf now seems impossible to cross over to once again stand on holy ground of Paradise.

As they sat there, repenting of their sin, the familiar radiant brilliance of their Creator emblazoned the evening skies. The divine brightness was far brighter than the setting sun. Their majestic Creator God, rose slowly above the high hills of Eden, and continued to ascend into Heaven. Sadly, they watched the glow of God, until He could no longer be seen. The man and his wife were now alone in the world, for they knew that their beloved LORD God would never return to walk with them in their beautiful garden, in the cool of the day. They were cursed humans, living in a cursed world that is filled with cursed fallen angels, who are led by the embodiment of all evil, Satan. They were now truly alone. All they have is each other.

Adam and Eve rested on the rock, consoling one another, and they began to recall all of the things that the Creator had taught, in the garden. Their hearts began to fill with joy and hope, as they remembered the promises of God and they rejoiced in the eternal hope of God's promises that reach to the very depths of the soul. Their LORD God may have returned to His Throne in Heaven, but His Word dwells deep in their hearts, and it can never change for God's Word is forever true. They take great comfort in knowing that they can always depend upon God's promises.

They recalled that God said, that their descendants would be as the countless stars of the heavens, and on one glorious appointed day, the Son of God shall leave the Throne of Heaven, and redeem all of those who believe and trust in Him. The Creator didn't tell them all of the details, but it was enough to know that the Creator God loves them so much that He will someday redeem them unto Himself.

Their hearts now fill with gladness and hope, and they vow to live each day for God. Adam takes Eve's hand, and they stand to their feet, and take one last look at Eden, and then turn and began walking toward their new world. Adam and Eve were rejoicing in their renewed faith, and boldly accepting the new adventures awaiting them. They were thrilled to know that they were the end of God's plan of *Creation*, and also the beginning of God's plan of *Redemption*. They rejoiced in knowing that the Redeemed will no longer be just the "companions" of the LORD God. When that glorious day of redemption will come, the Redeemed shall be His "counterpart," the loving Bride of the LORD God Almighty, to be forever joined together, for all eternity.

THE END…*of the beginning*

Epilogue

Creation and Redemption

We have traveled together back through time, and eternity, and beyond. Together, we have experienced God's plan of *Creation*, from its conception to its completion. We have explored concepts that address our very existence in the universe, and the universe itself. Men of science, philosophy, history, and theology, throughout the ages, have sought answers to these concepts. The concept of how the universe, and life upon this planet, came into existence, has always been the most intriguing topic of all of man's quests. Since no man has ever seen the creation of all that is, we have gone to the one source who states openly that He was there. The Creator God plainly states in His Word, the Holy Bible, that He alone created all things. In a few hundred words, Creator God tersely describes in detail, the entire plan of *"Creation"* of all that exists.

The story of Creation has taken us back beyond time itself. We have experienced a *"Journey Back to Eternity,"* a journey that has brought us into the presence of the radiant holy primeval light of God Almighty, when nothing else existed, but God. The primeval light of God is the source of all Light, for *"God is Light"* (1 John 1:5). The holy Light of

God, lit up all of Heaven, and made it sparkle like a jewel. That same Light set the universe ablaze in every star. The WORD said *"Let there be light, and there was light,"* and the whole Earth came alive with Light (Genesis 1:3). The true Light came into the world that He had made, but it knew Him not (John 9&10). The true Light, imputed to all His followers to be *"the light of the world"* (Matthew 5:14). In the New Jerusalem there will be no need for the light of the sun or moon, because the glory of God and the Lamb, shall be the Light of it (Revelation 21:23). The Light of God is eternal.

The WORD is forever part of the Triune God. God the Father gave the WORD, the Son of God, all authority and power to carry out their plan of *Creation and Redemption.* The WORD was the Creator of everything that exists (John 1:1-3). The WORD put aside all of the honor and glory in Heaven to become a man, *to become flesh and dwell among us* (John 1:14). The name of the WORD is found at the conclusion of the plan of *Redemption*, when the WORD, with the angels of God, and all of the "Redeemed" return to earth, at the end of the Tribulation, to reclaim the planet Earth from Satan. This is called, the "Revelation of Jesus Christ" (Revelation 21:11-16).

God has used in the Holy Scriptures, constant themes of the "Light of God," and the WORD of God, which we experienced in our search for truth. There is another theme that is, perhaps, the most important of all, that transcends the plan of *Creation and Redemption*, from eternity past to eternity future. There has been a crimson thread of "Blood" interwoven throughout the fabric of the history of man. We have seen the blood sacrifice in the Garden of Eden, which were the "lambs of creation," for the atonement of the sin of man. There were blood sacrifices of animals in the Tabernacle, and the Temples, for the sins of the nation of Israel, during the Dispensation of the Law. These were all pictures of the final sacrifice for sin; God's own Son, Jesus, the "Lamb of God," who was slain for the sins of the world. On an old rugged cross on a hill called Golgotha, the purchase price for sin was paid in full, for all eternity. God's plan for the *"Redemption"* of mankind is complete. We are now in the last days of the Dispensation of Grace, and the time to be redeemed is short.

To be redeemed, one must acknowledge that he is a sinner, and repent of his sins, and ask God to save him from eternal condemnation. Everyone who is to be saved must believe and trust that Jesus, the "Lamb of God," paid the penalty for sin on the cross, and rose from the dead, otherwise they will have to pay the eternal price of their own sins, forever in the Lake of Fire. There is no other way.

> *"For God sent not his Son into the world to condemn the world, but that the world through him might be saved.*
>
> *He that believeth on him is not condemned; but he that believeth not is condemned already, because he hath not believed in the name of the only begotten Son of God."*
>
> *(John 3:17&18)*

There is an old saying, "There is always room at the foot of the cross for one more." What more could God do to change our residence from the "Kingdom of Darkness," to the "Kingdom of God." The battle between good and evil rages on, even today, and you and I are part of that battle. Each person must choose which side he wants to be on, and where he will choose to spend eternity.

The Creator God only took three chapters, in the Bible, to relate His awesome plan of *Creation*, but He used the entire remaining Bible to unfold His plan of *Redemption.* It seems to me that God places less importance on the creation of all things, then He does on redeeming our souls. It cost the Creator God nothing to create the universe, but it cost dearly to redeem man. It cost God the Father, His beloved, only begotten Son, to die an agonizing, cruel, bloody death, upon the cross. God could offer nothing more precious to Himself, to purchase our salvation, so that we may have eternal life.

IMAGINATION AND REALITY

From Genesis to Revelation, we have utilized our imaginations, to be in the presence of the primeval light of Almighty God, when nothing else existed, to the eternal light of God and the Lamb, in the New Jerusalem. Guided by every word of God's account of the *"Creation,"* through our imaginations, we were there to see it all. We have experienced a *"Journey Back to Eternity,"* to witness the creation of Heaven, the angels of God, and the universe. Our trusty imaginary vehicle, the *"Eternity Explorer,"* has taken us to the rebellion in Heaven, and to the fall of the angels. We witnessed the recreation of planet Earth, and the six days of *"Creation."* The most powerful memories of the days of creation are of Adam and Eve in the Garden of Eden. In my imagination, it was so idyllic and beautiful.

Our wonderful, awesome experiences together were based on the reality of God's Word, expressed in our minds and imaginations. The reality of God's Word is truth, and we are able to access that truth through our imaginations. Perhaps, each of us now has a much better understanding of our reason for life, why we were created, and how we fit into Creator God's plan of *"Creation and Redemption."*

THE BEGINNING...*of the end*

"Imagination is more important than knowledge.

Without imagination, there would be no progress,

No hope of discovering what is over the next horizon."

Albert Einstein

Printed in the United States
5123